Inherently Safer Chemical Processes

Publications Available from the
CENTER FOR CHEMICAL PROCESS SAFETY
of the
AMERICAN INSTITUTE OF CHEMICAL ENGINEERS

Inherently Safer Chemical Processes

A LIFE CYCLE APPROACH

A CCPS CONCEPT BOOK

Robert E. Bollinger
David G. Clark
Arthur M. Dowell III
Rodger M. Ewbank
Dennis C. Hendershot
William K. Lutz
Steven I. Meszaros
Donald E. Park
Everett D. Wixom

Edited by
Daniel A. Crowl

CENTER FOR CHEMICAL PROCESS SAFETY (CCPS)
of the
AMERICAN INSTITUTE OF CHEMICAL ENGINEERS
345 East 47th Street, New York, New York 10017

Copyright © 1996
American Institute of Chemical Engineers
345 East 47th Street
New York, New York 10017

Library of Congress Cataloging-in Publication Data
Inherently safer processes : a life cycle approach / Robert
E. Bollinger . . . [et al.] : edited by Daniel A. Crowl.
     p.   cm. — (A CCPS concept book)
   Includes bibliographical references and index.
   ISBN 0–8169–0703–X
   1. Chemical engineering—safety measures.   I. Bollinger,
Robert E.   II. Crowl, Daniel A.   III. Series.
TP149.I47   1996
660′.2804—dc20   96–41370

Contents

Foreword

I had the good fortune to become involved in process safety at the right time. It was the right time for me personally as without my previous 16 years in production, I would not have had the right experience. It was also the right time in another sense as, in 1968, the subject was poised for take-off. The chemical industry was coming to realize that safety needed a technical input and was not just something that could be left to an elderly foreman.

There was much to do. What should we do first? My colleagues and I used what we now call quantitative risk assessment (QRA) to help us decide priorities—which risks to reduce first; which to leave alone for the time being. Our calculations were crude but within a couple of decades the technique had advanced so much that sophisticated computerized programs for estimating risk were available.

We put a lot of effort into improving safety by adding protective equipment onto our plants, new and old: gas detectors, emergency isolation valves, interlocks, steam curtains, fire insulation, catchment pits for LPG storage tanks, and so on. We also introduced new procedures, such as hazard and operability studies and modification control, or persuaded people to follow old ones, such as permits-to-work and audits.

According to Section 6.5 of this book, operators may need an hour or more to diagnose a fault. I am not a quick thinker and it took rather longer before a particular penny dropped. It was 1974, the time of Flixborough, before I realized that we ought, when possible, to be removing hazards rather than controlling them; that we ought to be reducing our inventories or using safer materials instead of toxic or flammable ones. Compared with QRA the idea was slow to catch on, but the last few years have seen a great growth of interest. The seed that took so long to germinate has now flowered and in this book has produced the finest bloom so far. The finest fruit will come when more people follow its advice and design and build more inherently safer plants—but the flowers have to come first.

This book is well written, logically developed, and easy to read. I hope it will be widely read, not just by designers, but by everyone involved with the design of new plants, including chemists who choose the reactions to be used. Above all, I hope it will be read by senior managers, as they are in a position to influence the policy and culture of the company and are inclined to ask why we carry out so many studies on new projects instead of getting on with the detailed design. This book will tell them why. I can think of no better Christmas present for your company president.

—*Trevor Kletz*

Preface

For over 30 years the American Institute of Chemical Engineers (AIChE) has been involved with process safety and loss control issues in the chemical, petrochemical, hydrocarbon process and related industries and facilities. AIChE publications and symposia are information resources for the chemical engineering and other professions on the causes of process incidents and the means of preventing their occurrences and mitigating their consequences.

The Center for Chemical Process Safety (CCPS), a Directorate of the AIChE, was established in 1985 to develop and disseminate technical information for use in the prevention of major chemical process incidents. With the support and direction of the CCPS Advisory and Managing Boards, a multifaceted program was established to address the need for Process Safety Management systems to reduce potential exposures to the public, the environment, personnel, and facilities. This program involves the development and publication of *Guidelines* relating to specific areas of Process Safety Management; organizing, convening and conducting seminars, symposia, training programs, and meetings on process safety-related matters; and cooperation with other organizations, both internationally and domestically, to promote process safety. Recently, CCPS has extended its publication program to include a "Concept Series" of books. These books are focused on more specific topics and are intended to complement the longer books in the *Guidelines* series. CCPS activities are supported by funding and professional expertise by over 90 corporations. Several government agencies and academic institutions also participate in CCPS endeavors.

In 1989, CCPS published the *Guidelines for Technical Management of Chemical Process Safety*, which presented a model for Process Safety Management characterized by twelve distinct, essential, and inter-related elements. The Foreword to that book stated:

> For the first time, all the essential elements and components of a model of a technical management program in chemical process safety have been assembled in one document. We believe the Guidelines provide the umbrella under which all other CCPS Technical Guidelines will be promulgated.

This "Concept Series" book, *Inherently Safer Chemical Processes: A Life Cycle Approach*, supports many of the twelve Key Elements of Process Safety, as identified in the *Guidelines for Technical Management of Chemical Process Safety*, including Capital Project Review and Design Procedures, Process Risk Management, Human Factors, Audits and Corrective Action, and Enhancement of Process Safety Knowledge. The purpose of this Concept Series book is to demonstrate the application of inherently safer strategies throughout all the stages of the chemical process life cycle.

Acknowledgments

Inherently Safer Chemical Processes—A Life Cycle Approach was written by the Center for Chemical Process Safety Inherently Safer Processes Subcommittee:

Arthur M. Dowell, III (Co-Chair), *Rohm and Haas Texas Inc.*
Donald E. Park (Co-Chair), *Albemarle Corporation*
Robert E. Bollinger, *Center for Chemical Process Safety*
David G. Clark, *E. I. DuPont de Nemours & Co.*
Rodger M. Ewbank, *Rhône-Poulenc, Inc.*
Dennis C. Hendershot, *Rohm and Haas Company*
William K. Lutz, *Union Carbide Corporation*
Steven I. Meszaros, *American Home Products Corp.*
Everett D. Wixom, *Exxon Chemical Company*

Dr. Daniel A. Crowl of Michigan Technological University served as editor for the book. The Subcommittee would also like to acknowledge the contributions of Mr. Robert A. Schulze of CCPS and Mr. Robert Kambach of BASF Corporation for their valuable input during the conceptual stages of this project.

The Subcommittee acknowledges the support and contributions of their employer organizations in completing this book. Mr. Bob G. Perry,

Dr. Jack Weaver, and Mr. Sanford Schreiber of CCPS sponsored and supported this project and provided access to the resources of CCPS and its sponsoring organizations. The Subcommittee particularly acknowledges the key contribution of Mr. Lester H. Wittenberg of CCPS in actively championing this project.

The authors thank the following for their assistance in word processing, creation of figures and tables, setting up committee meetings and teleconferences, and other administrative functions which were essential to the successful completion of this book: Ms. Christine Donahue, Rohm and Haas Company; Mrs. Dana Simmons, Albemarle Corporation; Ms. Christine Laraia, American Home Products Corporation; Ms. Nancy Marino, DuPont; Ms. Barbara Palmer, Exxon Chemical Company; and Ms. Carol Ann Talbert, Union Carbide Corporation. We particularly thank Ms. Donahue for receiving peer reviewer comments and distributing them to the authors.

Finally, the authors thank the following peer reviewers for taking the time to review the final manuscript:

Stanley E. Anderson,	*Rohm and Haas Texas Inc.*
Lisa Bendixen,	*Arthur D. Little, Inc.*
Fred Dyke,	*Arthur D. Little, Inc.*
Walter L. Frank,	*E. I. DuPont de Nemours & Company, Inc.*
Raymond A. Freeman,	*Monsanto Company*
Thomas Gibson,	*Dow Chemical Company*
John A. Hoffmeister,	*Lockheed Martin Energy Systems, Inc.*
Thomas Janicik,	*Solvay Polymers, Inc.*
Peter F. McGrath,	*Olin Corporation*
Larry Meier,	*The Hartford Steam Boiler Inspection and Insurance Company*
Joseph B. Mettalia, Jr.,	*Center for Chemical Process Safety*
Ken Murphy,	*United States Department of Energy*
Henry Ozog,	*Arthur D. Little, Inc.*
Gary Page,	*American Home Products Corporation*
James L. Paul,	*Hoechst Celanese Corporation*
Charles A. Pinkow,	*Rohm and Haas Company*
Harvey Rosenhouse,	*FMC Corporation*
John D. Snell,	*Occidental Chemical Corporation*
R. Peter Stickles,	*Arthur D. Little, Inc.*

Anthony A. Thompson,   *Monsanto Company*
William R. Tilton,   *E. I. DuPont de Nemours & Company, Inc.*
A. Sumner West,   *Center for Chemical Process Safety*
Jan Windhorst,   *Nova Chemicals, Ltd.*
Donald L. Winter,   *Mobil Oil Corporation*

1

Introduction

On December 14, 1977, Trevor Kletz, who was at that time safety advisor for the ICI Petrochemicals Division, presented the annual Jubilee Lecture to the Society of Chemical Industry in Widnes, England. His topic was "What you don't have, can't leak," and this lecture was the first clear and concise discussion of the concept of inherently safer chemical processes and plants.

Following the Flixborough explosion in 1974, there was increased interest in chemical process industry (CPI) safety, from within the industry as well as from government regulatory organizations and the general public. Much of the focus of this interest was on controlling the hazards associated with chemical processes and plants through improved procedures, additional safety interlocks and systems, and improved emergency response. Kletz proposed a different approach— to change the process to eliminate the hazard completely or reduce its magnitude sufficiently to eliminate the need for elaborate safety systems and procedures. Furthermore, this hazard elimination or reduction would be accomplished by means that were inherent in the process, and, thus, permanent and inseparable from it.

Kletz repeated the Jubilee Lecture twice in early 1978, and it was subsequently published (Kletz, 1978). In 1985 Kletz brought the concept of inherent safety to North America. His paper, "Inherently Safer Plants" (Kletz, 1985), won the Bill Doyle Award for the best

1

paper presented at the 19th Annual Loss Prevention Symposium, sponsored by the Safety and Health Division of the American Institute of Chemical Engineers.

Interest in inherently safer chemical processes and plants has grown over the years since 1978, and that growth has been particularly rapid in the 1990s (Kletz, 1996). In 1995 and 1996, there were more than 30 papers and presentations related to inherently safer chemical processes given at six different meetings, conferences, and congresses sponsored by the American Institute of Chemical Engineers and the Center for Chemical Process Safety. Inherently safer design is also receiving attention from government and regulatory organizations in the United States and Europe (Ashford, 1993; Lin, et al., 1994; Mansfield, 1994), joint industry-government working groups such as the INSIDE Project in Europe (Rogers et al., 1995; Mansfield, 1996), and environmental and public interest organizations (Tickner, 1994).

As the model in Figure 1.1 shows, process hazards come from two sources:

- hazards that are characteristic of the materials and chemistry used, and
- hazards that are characteristic of the process variables—the way the chemistry works in the process.

Hazards can be reduced or eliminated by changing the materials, chemistry, and process variables such that the reduced hazard is characteristic of the new conditions. The process with reduced hazards is described as inherently safer. This terminology recognizes there is no chemical process that is without risk, but all chemical processes can be made safer by applying inherently safer concepts. This book occasionally uses the term "inherent safety"; this does not mean absolute safety.

After the inherent hazards are reduced, layers of protection are frequently used to protect the receptors of the hazard—the public, the environment, workers, other processes, or the process itself (Figure 1.1). In the strictest sense, one could argue that the definition of inherently safer applies only to elimination or reduction of the hazard. In the broad sense, the strength of a layer of protection can be improved by features that are permanent and inseparable from that layer. Thus, layers of protection can be classified into three categories, listed in decreasing order of reliability: passive, active, and procedural. A passive layer of protection can be described as inherently safer than an active

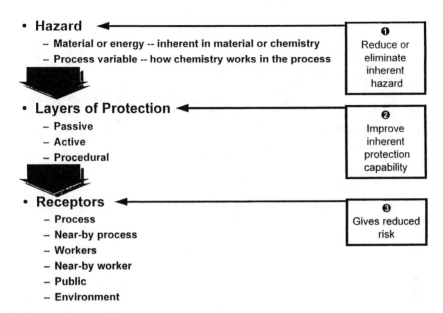

Figure 1.1. Reduce inherent hazards and improve inherent strength of protection layers to reduce risk

layer. Independent layers of protection are inherently safer than layers of protection that share common elements.

Inherently safer concepts will enhance overall risk management programs, whether directed toward reducing frequency or consequences of potential accidents. Inherently safer design strategies fall primarily in the inherent or passive categories, but can also be applied in the active or procedural categories. These strategies—minimize, substitute, moderate, and simplify—are discussed in detail in this book in Chapters 2 and 3.

The process industry has recognized that a chemical process goes through various stages of evolution. In this book, these stages are called *life cycle stages* as shown by Figure 1.2. The life cycle of a process begins with its discovery at the research stage. Then a process grows through stages of process development, design and construction, and matures with operations, maintenance, and modification. At the end, involvement with the process ends with decommissioning. Each section of Chapter 4 addresses a different life cycle stage and, thus, is aimed at a different audience.

Exploring inherently safer alternatives may require more resources during the early stages of development than is otherwise the case.

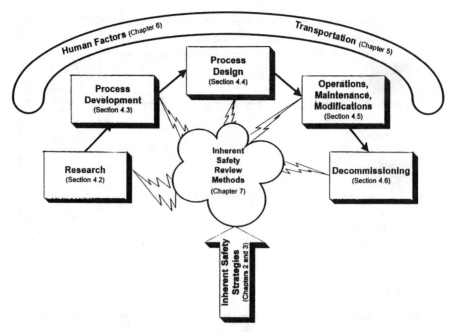

Figure 1.2. The process life cycle stages (with section references for this book).

However, the resulting understanding will, in many cases, minimize or eliminate the need for appended safety mitigation devices and the costs of maintaining them as well as reduce the possibility of incidents. Inherently safer considerations may reduce the life cycle cost of the process. In general, the economic benefits to be derived from inherently safer thinking will increase by application early in the process. However, it is never too late to start using inherently safer concepts for existing facilities.

Transportation should be considered when assessing risks associated with planned or existing plants. The design of new chemical processing units should include at the earliest opportunity a qualitative or quantitative risk assessment of the whole system including production, use, and transportation in order to minimize overall risk. A brief discussion of the inherent safety aspects of transportation is included in Chapter 5.

Human factors are an extremely important part of inherently safer concepts. Processes should be designed to avoid error traps. Chapter 6 of this book presents a discussion of human factors as related to inherently safer design.

Inherently safer is a way of thinking; one needs a fundamental understanding of the development and design processes for new plants or retrofit of existing plants as a single comprehensive system. Inherently safer thinking should be applied at the outset of the development process and continued throughout the life cycle of the process. Improved understanding of the process may result in a better process and higher quality products. Processes should be reviewed for hazards and risks periodically. Chapter 7 discusses review methods to do this.

Inherently safer as a way of thinking involves a holistic approach to consider the process as a system with interacting concerns such as toxicity, flammability, reactivity, stability, quality, capital, operating cost, transportation risks, and site factors. These considerations make it important to understand management of tradeoffs and definition of tolerable risks, be they safety, environmental, or financial in nature. Containment of energy, pressure, temperature, or chemicals will be an important consideration in many inherently safer suggestions, whereas there will be many constraints placed on the decision-making process due to economic limitations or perceived risks associated with new technology.

Process safety is fundamental to the basic practice of chemical engineering; thus, the concepts of inherently safer processes should be instilled in chemical engineering students at an early stage. Practicing engineers should be encouraged to adopt the concepts.

There is work to be done to improve the tools available for the application of inherently safer concepts. Chapter 8 discusses some opportunities for future research.

The objective of this book is to influence the future state of chemical process evolution by illustrating and emphasizing the merits of integrating process research, development, and design into a comprehensive process that balances safety, capital, and environmental concerns throughout the life cycle of the process. The authors hope that this book will influence the next generation of engineers and chemists as well as current practitioners and managers in the field of chemical processing.

2

Philosophy

2.1. A Way of Thinking

What do we mean when we speak of an "inherently safer" chemical process? "Inherent" has been defined as "existing in something as a permanent and inseparable element, quality, or attribute" (*American College Dictionary,* 1967). A chemical manufacturing process is **inherently safer** if it reduces or eliminates the hazards associated with materials and operations used in the process, and this reduction or elimination is permanent and inseparable. To appreciate this definition fully, it is essential to understand the precise meaning of the word "hazard." A hazard is defined as a physical or chemical characteristic that has the potential for causing harm to people, the environment, or property (adapted from CCPS, 1992). The key to this definition is that the hazard is intrinsic to the material, or to its conditions of storage or use. Some specific examples of hazards include:

- Chlorine is toxic by inhalation.
- Sulfuric acid is extremely corrosive to the skin.
- Ethylene is flammable.
- Steam confined in a drum at 600 psig contains a significant amount of potential energy (P_v energy).
- Acrylic acid can polymerize, releasing large amounts of heat.

7

These hazards cannot be changed—they are basic properties of the materials and the conditions of usage. The inherently safer approach is to reduce the hazard by reducing the quantity of hazardous material or energy, or by completely eliminating the hazardous agent.

Crowl and Louvar (1990) describe a three-step process which most accidents follow:

- *Initiation:* the event that starts the accident
- *Propagation:* the events that maintain or expand the accident
- *Termination:* the events that stop the accident or diminish it in size

Inherently safer strategies can impact the accident process at any of the three stages. The most effective strategies will prevent initiation of the accident. Inherently safer design can also reduce the potential for propagating an accident, or provide an early termination of the accident sequence before there are major impacts on people, property, or the environment.

One traditional risk management approach is to control the hazard by providing layers of protection between it and the people, property, and surrounding environment to be protected. These layers of protection may include operator supervision, control systems, alarms, interlocks, physical protection devices, and emergency response systems. Figure 2.1 illustrates the layers of protection concept, and includes examples of some layers which might be found in a typical chemical plant. This approach can be highly effective, and its application has resulted in significant improvement in the safety record of the chemical industry.

The approach of imposing barriers between a hazard and potentially impacted people, property, and environment has significant disadvantages:

- The layers of protection are expensive to build and maintain throughout the life of the process. Factors include initial capital expense, operating costs, safety training cost, maintenance cost, and diversion of scarce and valuable technical resources into maintenance and operation of the layers of protection.
- The hazard remains, and some combination of failures of the layers of protection may result in an accident. Since no layer of protection can be perfect, there is always some risk that an incident will occur.

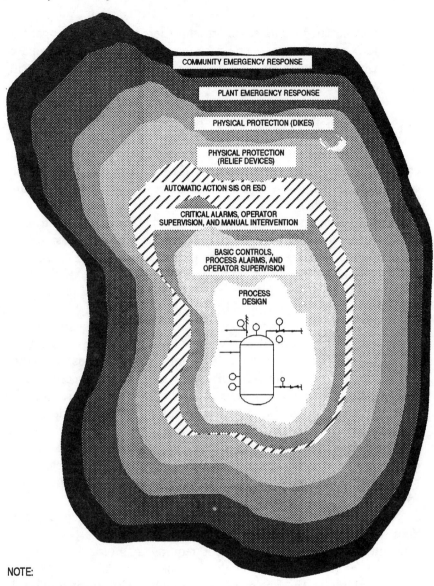

NOTE:

Protection layers for a typical process are shown in the order of activation expected as a hazardous condition is approached.

ESD – Emergency Shutdown
SIS – Safety Interlock System

Figure 2.1. Typical layers of protection in a modern chemical plant (CCPS 1993b).

- Because the hazard is still present, there is always a danger that its potential impacts could be realized by some unanticipated route or mechanism. Nature may be more creative in inventing ways by which a hazardous event can occur than experts are in identifying them. Accidents can occur by mechanisms that were unanticipated or poorly understood.

For these reasons, the inherently safer approach should be an essential aspect of any safety program. If the hazards can be eliminated or reduced, the extensive layers of protection to control those hazards will not be required.

There can be much discussion about whether or not a particular safety feature in a chemical process is "inherent." Such discussions may arise in part because different people are viewing the process at different levels of resolution, ranging from a global view of the entire process to a very detailed view of specific features of the process. The definition of hazard (an inherent physical or chemical characteristic that has the potential for causing harm to people, the environment, or property) can be applied at any level of resolution. For example, Joe, a process control engineer, describes an interlock system as "inherently safer" because it uses diverse multiple sensing elements, compared to an alternative design which uses multiple, but identical, sensors. Joe is looking at the process at a very detailed level, considering the characteristics of a layer of protection. The inherent physical characteristic of the system he defines as a hazard is the potential for common mode failure with identical sensing elements. Therefore, he regards the system using diverse sensing elements as inherently safer with regard to this very specific hazard.

On the other hand, Mary, a research process development engineer, does not consider Joe's system to be inherently safer, because a truly inherently safer system would not require an interlock at all. The process uses flammable materials and operates at elevated pressure. Mary, looking at the entire process, would only consider it to be inherently safer if the flammable materials were eliminated or the process was operated at ambient pressure. Mary is considering the inherent safety characteristics of the entire process, rather than a single interlock system.

From Joe's and Mary's viewpoints, each may be correct. Joe's diverse interlock system is indeed inherently safer as a layer of protection than the alternative using identical sensors, but it is still part of a process which is inherently less safe than alternatives which may be feasible. Mary is searching for that inherently safer process, which may make Joe's interlock system unnecessary. However, until she finds that alternative, it is shown to be feasible, and it is actually implemented, Joe's diverse interlock system is an inherently safer way of designing a required interlock. Both Joe and Mary are applying inherently safer concepts in their thinking about the process, but they are applying those concepts at different levels. Joe's interlock system does not represent true inherent safety for the process, but for purposes of this text, any improvement in a layer of protection which is permanent and inseparable, and not easily weakened or removed from the system, is considered to be a process safety improvement in an inherently safer direction.

In this book, we will adopt a broad definition of inherently safer. The discussion and examples range from basic process chemistry through details of the design of hardware and procedures. The authors recognize that some of these more detailed examples may arise in processes that would not be described as "inherently safer" overall. However, we believe that it is important to encourage inherently safer thinking at all levels of process and plant design, from the overall concept through the detailed design of equipment and procedures.

In considering the economics of process alternatives, it is important to think about the total life cycle costs. There is an increasing interest in this concept in the environmental area, with a recognition of the need to incorporate waste treatment, waste disposal, regulatory compliance, potential liability for environmental damage, and other long term environmental costs into project economic evaluation. Similarly, we must consider life cycle safety costs. Some examples of factors which should be considered include:

- Capital cost of safety and environmental equipment.
- Capital cost of passive barriers (for example, containment dikes, and vacant land to provide spacing, required by codes, regulations and insurers).
- Operating and maintenance costs for safety instruments and interlocks, fire protection systems, personal protective equipment, and other safety equipment.
- Increased maintenance cost for process equipment due to safety requirements (for example, safety permits, cleaning and purging equipment, personal protective equipment, training, and restricted access to process areas).

- Operator safety training costs for hazardous materials or processes.
- Regulatory compliance costs.
- Insurance costs.
- Potential property damage, product loss, and business interruption costs if an incident occurs.
- Potential liability if an incident occurs.

Inherently safer processes provide an opportunity to reduce or eliminate many of these long term economic costs. These benefits will not be realized unless we recognize and fully account for the long term costs associated with hazardous materials and processes.

Note that we refer to a process as "inherently safer," when compared to some other alternative, but not as "inherently safe." All materials and processes have hazards, and it is not realistic or practical to propose that we can eliminate all of them. In many cases we can identify material and process alternatives which clearly reduce hazards, and we can consider those alternatives to be inherently safer.

Inherently safer design is a fundamentally different way of thinking about the design of chemical processes and plants. It focuses on the elimination or reduction of the hazards, rather than on management and control. This approach should result in safer and more robust processes, and it is likely that these inherently safer processes will also be more economical in the long run (Kletz, 1984, 1991b).

2.2. The Role of Inherently Safer Design Concepts in Process Risk Management

How does inherently safer design fit into an overall process risk management program? To answer this question, it is first necessary to understand the definition of risk. Risk is defined as a measure of economic loss, human injury, or environmental damage in terms of both the incident likelihood and the magnitude of the loss, injury, or damage (CCPS, 1989a, 1995b). Any effort to reduce the risk arising from the operation of a chemical processing facility can be directed toward reducing the likelihood of incidents (incident frequency); reducing the magnitude of the loss, injury or damage should an incident occur (incident consequences), or some combination of both. In general, the strategy for reducing risk, whether directed toward reducing frequency or consequence of potential accidents, can be classified into four categories. These categories, in decreasing order of reliability, are:

- *Inherent*—Eliminating the hazard by using materials and process conditions which are nonhazardous; e.g., substituting water for a flammable solvent.
- *Passive*—Minimizing the hazard by process and equipment design features which reduce either the frequency or consequence of the hazard without the active functioning of any device; e.g., the use of equipment rated for higher pressure.
- *Active*—Using controls, safety interlocks, and emergency shutdown systems to detect and correct process deviations; e.g., a pump that is shut off by a high level switch in the downstream tank when the tank is 90% full. These systems are commonly referred to as engineering controls.
- *Procedural*—Using operating procedures, administrative checks, emergency response, and other management approaches to prevent incidents, or to minimize the effects of an incident; e.g., hot-work procedures and permits. These approaches are commonly referred to as administrative controls.

Risk control strategies in the first two categories, inherent and passive, are more reliable because they depend on the physical and chemical properties of the system rather than the successful operation of instruments, devices, procedures, and people. Inherent and passive strategies differ, but are often confused. A truly inherently safer process will reduce or completely eliminate the hazard (Kletz, 1991a), rather than simply reducing its impact. This book discusses both inherent and passive strategies to manage risk. Table 2.1 gives examples of the four risk management strategy categories. These categories are not rigidly defined, and some strategies may include aspects of more than one category.

There are also opportunities for making active and procedural risk management systems inherently safer. For example, consider two alternative designs for a high pressure interlock for a vessel:

1. A pressure sensor giving a continuous indication which is displayed on the control panel and can be observed by the operator. The sensor has a high pressure safety interlock set at a predetermined pressure that activates an emergency shutdown system.

TABLE 2.1

Examples of Process Risk Management Strategies (CCPS, 1993a)

Risk Management Strategy Category	Example	Comments
1. Inherent	An atmospheric pressure reaction using nonvolatile solvents which is incapable of generating any pressure in the event of a runaway reaction.	There is no potential for overpressure of the reactor because of the chemistry and physical properties of the materials.
2. Passive	A reaction capable of generating 150 psig pressure in case of a runaway, done in a 250 psig reactor.	The reactor can contain the runaway reaction. However, if 150 psig pressure is generated, the reactor could fail due to a defect, corrosion, physical damage or other cause.
3. Active	A reaction capable of generating 150 psig pressure in case of a runaway, done in a 15 psig reactor with a 5 psig high pressure interlock to stop reactant feeds and a properly sized 15 psig rupture disk discharging to an effluent treatment system.	The interlock could fail to stop the reaction in time, and the rupture disk could be plugged or improperly installed, resulting in reactor failure in case of a runaway reaction. The effluent treatment system could fail to prevent a hazardous release.
4. Procedural	The same reactor described in Example 3 above, but without the 5 psig high pressure interlock. Instead, the operator is instructed to monitor the reactor pressure and stop the reactant feeds if the pressure exceeds 5 psig.	There is a potential for human error, the operator failing to monitor the reactor pressure, or failing to stop the reactant feeds in time to prevent a runaway reaction.

Note: These examples refer only to the categorization of the risk management strategy with respect to the hazard of high pressure due to a runaway reaction. The processes described may involve trade-offs with other risks arising from other hazards. For example, the nonvolatile solvent in the first example may be extremely toxic, and the solvent in the remaining examples may be water. Decisions on process design must be based on a thorough evaluation of all of the hazards involved.

2. The same system, but with an on-off pressure switch set to activate the emergency shutdown system if the pressure reaches the predetermined point. The pressure switch remains inactive as long as the pressure is below its trip point.

Design alternative 1 is inherently safer because the pressure sensor provides continuous feedback to the operator. The operator has some confidence that the pressure sensor is working (although not complete assurance—it could be indicating incorrectly), and may observe that pressure is increasing before it reaches the high pressure trip point. However, both design alternatives are still classified as **active** systems. The first alternative is an inherently safer implementation of an active safety system. Section 2.6 discusses four strategies for identifying inherently safer process design. These **inherent safety strategies** can be applied to the implementation of any of the four **process risk management strategies**—inherent, passive, active, and procedural.

Marshall (1990, 1992) discusses accident prevention, control of occupational disease, and environmental protection in terms of strategic and tactical approaches. Strategic approaches have a wide significance and represent "once and for all" decisions. The inherent and passive categories of risk management would usually be classified as strategic approaches. In general, strategic approaches are best implemented at an early stage in the process or plant design. Tactical approaches include the active and procedural risk management categories. Tactical approaches tend to be implemented much later in the plant design process, or even after the plant is operating, and often involve much repetition, increasing the costs and potential for failure.

2.3. When Do We Consider Inherently Safer Options?

The search for inherently safer process options must begin early in the process life cycle, and never stop. The greatest potential opportunities for impacting process design occur early in the invention and development of processes. As stated by the National Research Council (Design, 1988), "Few basic decisions affect the hazard potential of a plant more than the initial choice of technology." Early in development there is a great deal of freedom in the selection of chemistry, solvents, raw materials, process intermediates, unit operations, plant location, and other process parameters. As the process moves through its life cycle, it becomes more difficult and expensive to change the basic process.

> **The two most important tools an architect has are the eraser in the drawing room and the sledge hammer on the construction site.**
>
> —Frank Lloyd Wright
>
> **Which would you rather use?**

It is never too late to consider inherently safer alternatives. Major enhancements to the inherent safety of plants which have been operating for many years have been reported (CCPS, 1993a; Wade, 1987; Carrithers et al., 1996)

Table 2.2 is a summary of some of the key questions that should be asked at various stages in the development of a plant design, as suggested by the INSIDE Project, a major European government/

TABLE 2.2

Some Key Questions and Decisions Related to Inherently Safer Plant Design
(Rogers et al., 1995)

Decision Point	Key Questions and Decisions	Information Used
Initial Specification	What product? What capacity?	Market research Research and Development new product ideas
Process Synthesis Route	How to make the product? What route? What reactions, materials, starting points?	Research and Development chemists research Known synthesis routes and techniques
Chemical Flowsheet	Basic unit operation selection with flow rates, conversion factors, temperatures, pressures, solvents and catalyst selection	Process synthesis route Laboratory and pilot scale trials Knowledge of existing processes
Process Flowsheet	Batch vs. Continuous operation Detailed unit operations selection Control and operation philosophy	Information above plus process engineering design principles and experience
Process Conceptual Design	Equipment selection and sizing Inventory of process Single vs. Multiple trains Utility requirements Overdesign and flexibility Recycles and buffer capacities Instrumentation and control Location of plant Preliminary plant layout Materials of construction	As above plus equipment suppliers data, raw materials data, company design procedures and requirements
Process Detailed Design	Detailed specification based on concept design Minimize number of possible leak paths Make plant "friendly" to control, operate, and maintain Avoid or simplify hazardous activities such as sampling, loading and unloading	Process conceptual design and codes/standards and procedures Experience on past projects/designs

industry project sponsored by the Commission of the European Community (Rogers et al., 1995; Mansfield, 1996). The INSIDE project reviews the current status of inherently safer process and plant design, and is developing approaches to encourage further application of inherently safer design philosophies. Chapter 4 of this book also provides additional guidance on the application of inherently safer design principles at various stages in a project life cycle.

To be most effective, an inherent safety program should have the objective of creating an awareness of inherent safety in a broad range of chemists and engineers involved in the development of products and processes throughout an organization. The inherently safer way of doing things should become a way of thinking and working. Perhaps the critical moment in the life of any idea occurs immediately after the idea springs into a person's head. Does he or she pursue the idea, file it away for further thought, discuss it with a colleague, or just forget it as being of no further interest? If everyone in the organization understands that inherently safer products and processes are valued and desired, it is much more likely that ideas with the potential to develop into inherently safer systems will survive this critical moment, and will grow and mature.

2.4 Inherent Safety Tradeoffs

In many cases, the inherent safety advantages of one process are clear when compared with alternatives. One or more hazards may be significantly reduced, while others are unaffected or only marginally increased. For example, aqueous latex paints are clearly inherently safer than solvent based paints, although there are applications where the increased performance of solvent based paints justifies their use, with the appropriate layers of protection.

Unfortunately, many times it is not clear which of several process alternatives is inherently safer. Because nearly all chemical processes have a number of hazards associated with them, an alternative which reduces one hazard may increase a different hazard. For example, process A uses flammable materials of low toxicity; process B uses noncombustible materials, which are volatile and moderately toxic, and process C uses noncombustible and nontoxic materials but operates at high pressure. Which process is inherently safer? The answer to this question will depend on the specific details of the process options. In addressing these questions, it is essential that *all* hazards be identified and understood, including:

- Acute toxicity
- Chronic toxicity
- Flammability
- Reactivity
- Instability
- Extreme conditions (temperature or pressure)
- Environmental hazards, including
  —Air pollution
  —Water pollution
  —Groundwater contamination
  —Waste disposal

The hazards associated with normal plant operations, such as normal stack emissions and fugitive emissions, as well as those resulting from specific incidents such as spills, leaks, fires and explosions should be considered.

It is also necessary to consider business and economic factors in making a process selection. These include:

- Capital investment
- Product quality
- Total manufacturing costs
- Operability of the plant
- Demolition and future clean-up and disposal cost

Design strategies which result in an inherently safer design may also tend to improve process economics. For example, minimizing the size of equipment or simplifying a process by eliminating equipment will usually reduce capital investment and reduce operating costs. However, overall process economics are very complex and are impacted by many factors, and it may not always be true that an inherently safer process is also economically more attractive.

An inherently safer process offers greater safety potential, often at a lower cost. However, selection of an inherently safer technology does not guarantee that the actual implementation of that technology will result in a safer operation than an alternate process which is inherently safer. The traditional strategy of providing layers of protection for an inherently more hazardous process can be quite effective, although the expenditure of resources to install and maintain the layers of protection may be very large. In some cases the benefits of the inherently more hazardous technology will be sufficient to justify the costs needed to provide the layers of protection required to reduce the risk to a tolerable level. As an example, Hendershot (1995b) compares the inherent safety characteristics of air and automobile transportation, and concludes that automobile transportation is inherently safer for reasons such as:

- The automobile, on the ground, will coast to a stop in case of engine failure, while the airplane will rapidly descend and may not be able to land safely.
- The automobile travels at a lower speed.
- The automobile contains a smaller inventory of passengers.
- The control of an automobile is simpler (in two dimensions) compared to the airplane, which must be controlled in three dimensions.

However, the benefits of air transportation, primarily speed, make it an attractive alternative for longer trips. These benefits have justified the expenditure of large amounts of money for providing extensive layers of protection to overcome the inherent hazards of air travel. The result is that air travel, while inherently more hazardous, is in fact safer than automobile travel for long trips. Similar situations can be expected to occur in the chemical process industry.

However, even when we determine that the benefits of an inherently less safe technology justify its use, we should always continue to look for inherently safer alternatives. Technology continues to evolve and advance, and inherently safer alternatives which are not economically attractive today may be very attractive in the future. The development of new inherently safer technology offers the promise of more reliably and economically meeting process safety goals.

McQuaid (1991), CCPS (1993a), and Hendershot (1995a) review a number of specific examples of inherent safety tradeoffs. These include:

- Chlorofluorocarbon (CFC) refrigerants are inherently safer with respect to fire, explosion, and acute toxic hazards when compared to alternative refrigerants such as ammonia, propane, and sulfur dioxide. However, they are believed to cause long term environmental damage because of stratospheric ozone depletion.
- Supercritical processing may use relatively nonhazardous materials such as water or carbon dioxide as reaction and extraction

solvents, but supercritical processing requires high temperatures and pressures.

- A solvent used in an exothermic reaction is nonvolatile, and
  moderately toxic. An alternative solvent is less toxic, but also
  has a much lower boiling point. There is a trade-off between
  toxic hazards and the potential for tempering the exotherm, but
  also generating pressure from boiling solvent in case of a runaway reaction.

2.5. Resolving Inherent Safety Issues

Deciding among a number of process options having inherent safety
advantages and disadvantages with respect to different hazards can be
quite difficult. The first step is to understand thoroughly all hazards
associated with the process options. Process hazard analysis and evaluation techniques are appropriate tools (CCPS, 1992). These include:

- past history and experience
- interaction matrices
- what if
- checklists
- what if/checklists
- hazard and operability (HAZOP) studies

The hazard identification step is perhaps the most important,
because any hazard not identified will not be considered in the decision
process. For example, the impact of chlorofluorocarbons on atmospheric ozone was unknown for much of the period of their use, and
this potential hazard was not considered until recent years.

Once the hazards have been identified, the process options can be
ranked in terms of inherent safety with respect to all identified hazards.
This ranking can be qualitative, placing hazards into consequence and
likelihood categories based on experience and engineering judgment
(CCPS, 1992). More quantitative systems can also be used to rank
certain specific types of hazard, for example, the Dow Fire and
Explosion Index (Dow, 1994b; Gowland, 1996a,b) and the Dow Chemical Exposure Index (Dow, 1994a). Unfortunately, none of these indices
consider the full range of hazards. To get an overall assessment of the
process options, it is necessary to use a variety of indices and qualitative
techniques and then combine the results. Initial work on an overall

inherent safety index which considers a wide variety of different hazards is being done at Loughborough University in the United Kingdom (Edwards et al., 1996).

Sometimes the consequences of all hazardous incidents can be expressed by a single common measure—for example, dollar value of property damage, total economic loss, risk of immediate fatality due to fire, explosion or toxic material exposure. If all consequences can be measured on a common scale, the techniques of quantitative risk analysis (CCPS, 1989a, 1995b) may be useful in assessing the relative magnitude of various hazards, and in understanding and ranking total risk of process options.

> **The true rule, in determining to embrace, or reject any thing, is not whether it have any evil in it; but whether it have more of evil, than of good. There are few things wholly evil, or wholly good. Almost every thing . . . is an inseparable compound of the two; so that our best judgment of the preponderance between them is continually demanded.**
>
> —Abraham Lincoln

In many cases, it is not readily apparent how the potential impacts from different hazards can be translated into some common scale or measure. For example, how do you compare long term environmental damage and health risks from use of CFC refrigerants to the immediate risk of fatality from the fire, explosion, and toxicity hazards associated with many alternative refrigerants? This question does not have a "right" answer. It is not really a scientific question, but instead it is a question of values. Individuals, companies, and society must determine how to value different kinds of risks relative to each other, and base decisions on this evaluation.

In many cases, formal tools for decision making can be useful, particularly if the hazards vary greatly in type of consequence or impact. Many of these tools introduce additional rigor, consistency, and logic into the decision process. Some available methods include:

- Weighted scoring methods, such as Kepner-Tregoe Decision Analysis and the Analytical Hierarchy Process (AHP)

- Cost–benefit analysis
- Payoff matrix analysis
- Decision analysis
- Multiattribute utility analysis

CCPS (1995a) reviews these decision aids and others, with special emphasis on how they are employed in making chemical process safety decisions. The chemical process industry is beginning to use these techniques in making safety, health, and environmental decisions. A weighted scoring technique based on Kepner–Tregoe Decision Analysis (Kepner and Tregoe, 1981) has been used (Hendershot, 1995a, 1996), as illustrated for a generic process in Table 2.3. Reid and Christensen (1994) describe the use of the Analytical Hierarchy Process to evaluate three alternative technologies considered for a metal fabrication application, with the overall goal of minimizing waste from the process.

2.6. Inherently Safer Design Strategies

Approaches to the design of inherently safer processes and plants have been grouped into four major strategies by IChemE and IPSG (1995) and Kletz (1984, 1991b):

Minimize	Use smaller quantities of hazardous substances (also called *Intensification*)
Substitute	Replace a material with a less hazardous substance
Moderate	Use less hazardous conditions, a less hazardous form of a material, or facilities which minimize the impact of a release of hazardous material or energy (also called *Attenuation* and *Limitation of Effects*).
Simplify	Design facilities which eliminate unnecessary complexity and make operating errors less likely, and which are forgiving of errors which are made (also called *Error Tolerance*).

These inherently safer design strategies are discussed in more detail in Chapter 3, and examples can also be found in Chapter 4, which discusses inherently safer design opportunities through the life cycle of a chemical process.

TABLE 2.3

An Example of a Weighted Scoring Decision Matrix

Parameter	Weighting Factors	Process Options			
		#1	**#2**	**#3**	**#4**
COST	*Performance Factor >*	× 2 =	× 9 =	× 10 =	× 1 =
	9	18	81	90	9
SAFETY	*Performance Factor >*	× 10 =	× 5 =	× 3 =	× 1 =
	10	100	50	30	10
ENVIRONMENT	*Performance Factor >*	× 3 =	× 5 =	× 1 =	× 10 =
	7	21	35	7	70
OPERABILITY	*Performance Factor >*	× 3 =	× 10 =	× 2 =	× 1 =
	5	15	50	10	5
DESIGN	*Performance Factor >*	× 1 =	× 9 =	× 10 =	× 3 =
	3	3	27	30	9
OTHER	*Performance Factor >*	× 7 =	× 5 =	× 10 =	× 1 =
	3	21	15	30	3
	SUM	178	258	197	106

PROCEDURE:

1. Assign a weighting factor (1 to 10) to the various parameters based on your judgment of the relative importance of this Safety, Health and Environmental, or other issue.

2. For each option, assign a performance factor from 1 to 10 based on the relative performance of that option with respect to the particular parameter. This can be based on judgment, or scaled based on some kind of quantitative analysis.

3. Multiply the weighting factor by the performance factor for each parameter and process option combination.

4. Sum the products for each process option.

The highest total is most desirable.

NOTES:

1. See Kepner and Tregoe (1981) or CCPS (1995a) for additional discussion, particularly on how potential negative consequences may impact the scoring matrix.

2. The weighting factors in this table are for purposes of illustrating the methodology only, and do not represent recommendations on the relative importance of the factors listed.

2.7. Summary

Inherently safer design represents a fundamentally different approach to chemical process safety. Rather than accepting the hazards in a process, and then adding on safety systems and layers of protection to control those hazards, the process designer is challenged to reconsider the process and eliminate the hazards. If the designer cannot eliminate the hazards, the challenge becomes to minimize or reduce them as much as possible by modifying the process, rather than by adding external layers of protection.

Process risk management strategies can be categorized, in order of decreasing reliability, as inherent, passive, active, and procedural. Inherently safer design strategies can be applied to any of these risk management strategies, as illustrated in Figure 2.2. It is possible, for example, to describe one procedure as inherently safer than another, perhaps because the "inherently safer" procedure is simpler. However, both procedures would still represent procedural approaches to process risk management.

Finally, in considering inherently safer design alternatives, it is essential to remember that there are often, perhaps always, conflicting

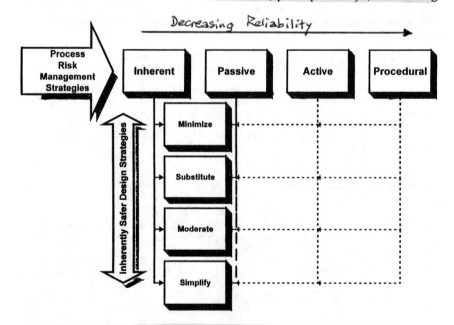

Figure 2.2. The relationship between process risk management and inherently safer design strategies.

benefits and deficiencies associated with the different options. Chemical processes usually have many potential hazards, and a change which reduces one hazard may create a new one or increase the magnitude of a different existing hazard. It is essential that the process designer retain a broad overview of the process when considering alternatives, that he/she remains aware of **all** hazards associated with each process option, and that appropriate tools are applied to chose the overall best option.

3

Inherently Safer
Design Strategies

Chapter 2 introduced four strategies for inherently safer design:

- Minimize
- Substitute
- Moderate
- Simplify

This chapter provides examples of these strategies, and the process life cycle discussions in Chapters 4 through 6 include additional examples.

3.1. Minimize

To minimize is to reduce the quantity of material or energy contained in a manufacturing process or plant. We often think of process minimization as resulting from the application of innovative new technology to a chemical process, for example, tubular reactors with static mixing elements, centrifugal distillation techniques, or innovative, high surface area heat exchangers. However, we must not forget that much can be accomplished in process inventory reduction simply by applying good engineering design principles with more conventional technology. Application of reliability-centered maintenance techniques can also increase the inherent safety of a plant by reducing plant downtime, thus reducing the need for intermediate inventory and storage. This in-process storage or surge capacity may be required to allow portions of the plant to continue to operate while other parts of the plant are shut down because equipment requires maintenance. Improving the reliability of critical pieces of equipment may eliminate or significantly reduce the need for in-process storage of hazardous chemical intermediates.

When designing a plant, every piece of process equipment should be specified as large enough to do its job, and no larger. We should minimize the size of all raw material and in-process intermediate storage tanks, and question the need for all in-process inventories, particularly of hazardous materials. Minimizing the size of equipment not only enhances inherent process safety, but it can often save money.

In 1877 Arthur M. Wellington, in his book *The Economic Theory of the Location of Railroads*, published a famous definition of engineering (Petroski, 1995):

> It would be well if engineering were less generally thought of, and even defined, as the art of constructing. In a certain important sense it is rather the art of not constructing; or to define it rudely, but not ineptly, it is the art of doing well with one dollar, which any bungler can do with two after a fashion.

If we can eliminate equipment from a manufacturing process, we do not have to design, purchase, operate, or maintain that equipment, thus saving money. Equipment which is eliminated also cannot leak or otherwise release hazardous material or energy into the surrounding environment. The true art of the engineer is to determine how to accomplish a given task with a minimum of equipment, and with the required equipment of the smallest size. Siirola (1995) discusses process synthesis strategies which are helpful in designing and optimizing a process route to minimize the equipment and operations required.

The term "process intensification" is used synonymously with "minimization." "Process intensification" is also often used more specifically to describe new technologies which reduce the size of unit operations equipment, particularly reactors. Innovative process intensification techniques are receiving more and more attention. Interesting possibilities for a range of unit operations, including reaction, gas–liquid contacting, liquid–liquid separation, heat exchange, distillation, and separation were reviewed during an international conference on process intensification (Akay and Azzopardi, 1995). Although the focus of this conference was on improving process economics, many of the technologies described have the potential for improving the inherent safety of processes as well, by virtue of the reduction of in-process inventories resulting from their application.

Benson and Ponton (1993) and Ponton (1996) have speculated on the ultimate results of continuing efforts for process minimization. They envision a twenty-first century chemical industry totally revolutionized by technological innovation, automation, and miniaturization. Small, distributed manufacturing facilities would produce materials on demand, at the location where they are needed. Raw materials would be nonhazardous, and the manufacturing processes would be waste free and inherently safe. While their vision of future technology is speculative, we are beginning to see progress in this direction.

A few examples of process minimization will be presented here. Kletz (1984, 1991b), Englund (1990, 1991a,b, 1993), IChemE and IPSG (1995), Lutz (1995a, b) and CCPS (1993a) provide many more examples.

Reactors

Reactors can represent a large portion of the risk in a chemical process. A complete understanding of reaction mechanism and kinetics is essential to the optimal design of a reactor system. This includes both the chemical reactions and mechanisms, as well as physical factors such as mass transfer, heat transfer, and mixing. A reactor may be large because the chemical reaction is slow. However, in many cases the chemical reaction actually occurs very quickly, but it appears to be slow due to inadequate mixing and contacting of the reactants. Innovative reactor designs which improve mixing may result in much smaller reactors. Such designs are usually cheaper to build and operate, as well as being safer due to smaller inventory. In many cases, improved product quality and yield also result from better and more uniform contacting of reactants. With a thorough understanding of the reaction, the designer can identify reactor configurations that maximize yield and minimize size, resulting in a more economical process, reducing generation of by-products and waste, and increasing inherent safety by reducing the reactor size and inventories of all materials.

Continuous Stirred Tank Reactors

A continuous stirred tank reactor (CSTR) is usually much smaller than a batch reactor for a specific production rate. In addition to reduced inventory, using a CSTR usually results in other benefits which enhance safety, reduce costs, and improve the product quality. For example:

- Mixing in the smaller reactor is generally better. Improved mixing may improve product uniformity and reduce by-product formation.
- Controlling temperature is easier and the risk of thermal runaway is reduced. Greater heat transfer surface per unit of reactor volume is provided by a smaller reactor.
- Containing a runaway reaction is more practical by building a smaller but stronger reactor rated for higher pressure.

In considering the relative safety of batch and continuous processing, it is important to fully understand any differences in chemistry and processing conditions, which may outweigh the benefits of reduced size of a continuous reactor.. Englund (1991b) describes continuous latex processes which have enough unreacted monomer in the continuous reactor that they are less safe than a well designed batch process.

Tubular Reactors

Tubular reactors often offer the greatest potential for inventory reduction. They are usually extremely simple in design, containing no moving parts and a minimum number of joints and connections. A relatively slow reaction can be completed in a long tubular reactor if mixing is adequate. There are many devices available for providing mixing in tubular reactors, including jet mixers, eductors, and static mixers.

It is generally desirable to minimize the diameter of a tubular reactor, because the leak rate in case of a tube failure is proportional to its cross-sectional area. For exothermic reactions, heat transfer will also be more efficient with a smaller tubular reactor. However, these advantages must be balanced against the higher pressure drop due to flow through smaller reactor tubes.

Loop Reactors

A loop reactor is a continuous steel tube or pipe which connects the outlet of a circulation pump to its inlet (Figure 3.1). Reactants are fed

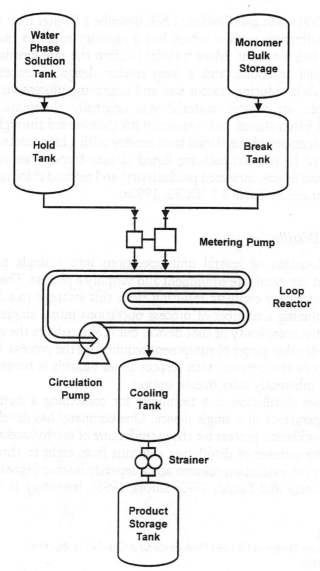

Figure 3.1. A loop reactor production system (Wilkinson and Geddes, 1993).

into the loop, where the reaction occurs, and product is withdrawn from the loop. Loop reactors have been used in place of batch stirred tank reactors in a variety of applications including chlorination, ethoxylation, hydrogenation, and polymerization. A loop reactor is typically much smaller than a batch reactor producing the same amount of product. Wilkinson and Geddes (1993) describe a 50-liter loop reactor for polymerization process which has a capacity equal to that of a 5000-liter batch reactor. Mass transfer is often the rate limiting step in gas–liquid reactions, and a loop reactor design increases mass transfer, while reducing reactor size and improving process yields. As an example, an organic material was originally chlorinated in a glass-lined batch stirred tank reactor, with chlorine fed through a dip pipe. Replacement of the stirred tank reactor with a loop reactor, with chlorine fed to the recirculating liquid stream through an eductor, reduced reactor size, increased productivity, and reduced chlorine usage as summarized in Table 3.1 (CCPS, 1993a).

Reactive Distillation

The combination of several unit operations into a single piece of equipment can eliminate equipment and simplify a process. There may be inherent safety conflicts resulting from this strategy (see Section 2.4). Combining a number of process operations into a single device increases the complexity of that device, but it also reduces the number of vessels or other pieces of equipment required for the process. Careful evaluation of the options with respect to all hazards is necessary to select the inherently safer overall option.

Reactive distillation is a technique for combining a number of process operations in a single device. One company has developed a reactive distillation process for the manufacture of methyl acetate that reduces the number of distillation columns from eight to three, also eliminating an extraction column and a separate reactor (Agreda et al., 1990; Doherty and Buzad, 1992; Siirola, 1995). Inventory is reduced

TABLE 3.1

Effect of Reactor Design on Size and Productivity for a Gas–Liquid Reaction (CCPS, 1993a)

Reactor Type	Batch Stirred Tank Reactor	Loop Reactor
Reactor size (l)	8000	2500
Chlorination time (hr)	16	4
Productivity (kg/hr)	370	530
Chlorine usage (kg/100 kg product)	33	22
Caustic usage in vent scrubber (kg/100 kg product)	31	5

and auxiliary equipment such as reboilers, condensers, pumps, and heat exchangers are eliminated. Figure 3.2 shows the conventional design, and Figure 3.3 shows the reactive distillation design. Siirola (1995) reports significant reductions in both capital investment and operating cost for the reactive distillation process.

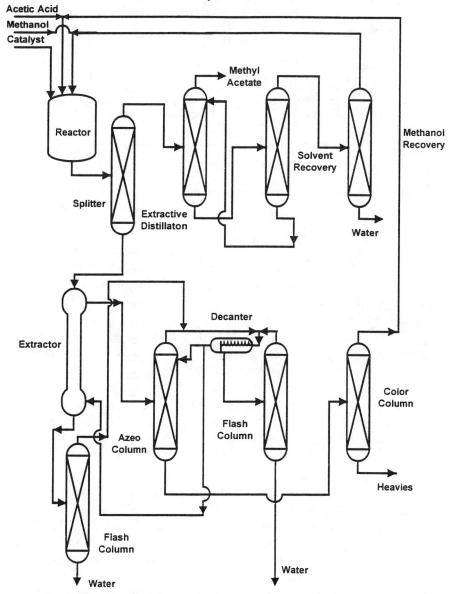

Figure 3.2. Conventional process for methyl acetate (based on Siirola, 1995).

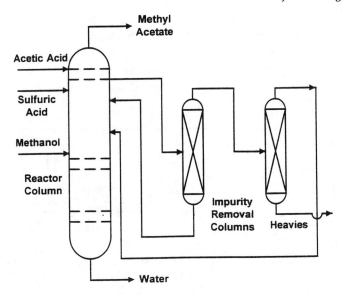

Figure 3.3. Reactive distillation methyl acetate process (based on Agreda et al., 1990).

Storage and Material Transfer

Raw material and in-process storage tanks and pipelines often represent a major portion of the risk of a chemical plant. Attention to the design of storage and transfer equipment can reduce hazardous material inventory.

Storage tanks for raw materials and intermediates are often much larger than really necessary, usually because this makes it "easier" to operate the plant. The operating staff can pay less attention to ordering raw materials on time, or can accept downtime in a downstream processing unit because upstream production can be kept in storage until the downstream unit is back on line. This convenience in operation can come at a significant cost in the risk of loss of containment of the hazardous materials being stored. The process design engineers and operating staff must jointly determine the need for all intermediate hazardous material storage, and minimize quantities where appropriate.

Similarly, hazardous raw material storage should also be minimized, with greater attention being given to "just in time" supply. Inventory reduction lowers inventory costs, while increasing inherent safety. In determining appropriate raw material inventories, the entire raw material supply chain must be considered. Will the supplying plant have to increase inventories to provide "just in time" service, and will this represent a greater risk than a larger inventory at the user facility? Will the raw material be stockpiled in a local storage facility, or in parked railroad cars or tank trucks, perhaps at a greater risk than on-site storage in a well designed facility? How much additional burden will "just-in-time" delivery place on operating staff? Will unplanned shutdowns due to running out of raw materials increase risks? Chapter 5 discusses inherently safer options in material transportation in more detail.

The reduction in inventory resulting from greater attention to plant operations and design of unit interactions can be substantial. Wade (1987) gives several excellent examples.

- An acrylonitrile plant eliminated 500,000 pounds of in-process storage of hydrogen cyanide by accepting a shutdown of the entire unit when the product purification area shut down. This forced the plant staff to solve the problems which caused the purification area shutdowns.
- Another acrylonitrile plant supplied by-product hydrogen cyanide to various other units. An inventory of 350,000 pounds of hydrogen cyanide was eliminated by having the other units draw directly from the acrylonitrile plant. This required considerable work to resolve many issues related to acrylonitrile purity and unit scheduling.
- A central bulk chlorine system with large storage tanks and extensive piping was replaced with a number of small cylinder facilities local to the individual chlorine users. Total inventory of chlorine was reduced by over 100,000 pounds. This is another example of conflicting inherent safety strategies. Use of the central bulk chlorine system reduces the need for operators to connect and disconnect chlorine cylinders, but with the disadvantage of a large inventory which could be released if a leak occurs. The use of a number of local cylinder facilities results in a greater likelihood of a leak because of the necessity to connect and disconnect the cylinders more frequently—but the maximum size of the leak will be limited to the inventory in one cylinder.

Other Examples

Section 4.3 on Process Development includes additional examples of opportunities for process minimization or intensification, including:

- Process piping
- Distillation
- Heat transfer equipment

3.2 Substitute

Substitution means the replacement of a hazardous material or process with an alternative which reduces or eliminates the hazard. Process designers, line managers, and plant technical staff should continually ask if less hazardous alternatives can be effectively substituted for all hazardous materials used in a manufacturing process. Examples of substitution in two categories are discussed—reaction chemistry and solvent usage. There are many other areas where opportunities for substitution of less hazardous materials can be found, for example, materials of construction, heat transfer media, insulation, and shipping containers.

Reaction Chemistry

Basic process chemistry using less hazardous materials and chemical reactions offers the greatest potential for improving inherent safety in the chemical industry. Alternate chemistry may use less hazardous raw material or intermediates, reduced inventories of hazardous materials, or less severe processing conditions. Identification of catalysts to enhance reaction selectivity or to allow desired reactions to be carried out at a lower temperature or pressure is often a key to development of inherently safer chemical synthesis routes. Some specific examples of innovations in process chemistry which result in inherently safer processes include:

- The insecticide carbaryl can be produced by several routes, some of which do not use methyl isocyanate, or which generate only small quantities of this toxic material as an in-process intermediate (Kletz, 1991b). One company has developed a proprietary process for manufacture of carbamate insecticides which generates methyl isocyanate as an in-situ intermediate. Total methyl isocyanate inventory in the process is no more than 10 kilograms (Kharbanda and Stallworthy, 1988; Manzer, 1994).

- Acrylonitrile can be manufactured by reacting acetylene with hydrogen cyanide:

$$CH{\equiv}CH + HCN \rightarrow CH_2{=}CHCN$$

Acetylene   Hydrogen   Acrylonitrile
          cyanide

A new ammoxidation process uses less hazardous raw materials (propylene and ammonia (Dale, 1987; Puranik et al., 1990).

$$CH_2{=}CHCH_3 + NH_3 + \tfrac{3}{2}O_2 \rightarrow CH_2{=}CHCN + 3H_2O$$

Propylene   Ammonia   Acrylonitrile

This process does produce HCN as a by-product in small quantities. Puranik et al. (1990) report on work to develop an improved, more selective catalyst, and on coupling the ammoxidation process with a second reactor in which a subsequent oxycyanation reaction would convert the by-product HCN to acrylonitrile.

- The Reppe process for manufacture of acrylic esters uses acetylene and carbon monoxide, with a nickel carbonyl catalyst having high acute and longterm toxicity, to react with an alcohol to make the corresponding acrylic ester:

$$CH{\equiv}CH + CO + ROH \xrightarrow[HCl]{Ni(CO)_4} CH_2{=}CHCO_2R$$

Acetylene   Alcohol   Acrylic ester

The new propylene oxidation process uses less hazardous materials to manufacture acrylic acid, followed by esterification with the appropriate alcohol (Hochheiser, 1986):

$$CH_2{=}CHCH_3 + \tfrac{3}{2}O_2 \xrightarrow{catalyst} CH_2{=}CHCO_2H + H_2O$$

Propylene   Acrylic acid

$$CH_2{=}CHCO_2H + ROH \xrightarrow{H^+} CH_2{=}CHCO_2R + H_2O$$

Acrylic acid   Alcohol   Acrylic ester

- Polymer supported reagents, catalysts, protecting groups, and mediators can be used in place of the corresponding small molecule materials (Sherrington, 1991; Sundell and Nasman, 1993). The reactive species is tightly bound to a macromolecular support which immobilizes it. This generally makes toxic, noxious, or corrosive materials much safer. The use of polystyrene sulfonic acid catalyst for the manufacture of methyl *t*-butyl ether (MTBE) from methanol and isobutene (isobutylene) is one example in commercial use.

$$CH_3OH + CH_2=C(CH_3)_2 \xrightarrow{\text{polystyrene sulfonic acid}} CH_3=OC(CH_3)_3$$

Methanol   Isobutane   MTBE

- The chemistry of side reactions and by-products may also offer opportunities for increasing the inherent safety of a process. For example, a process involving a caustic hydrolysis step uses ethylene dichloride (EDC; 1,2-dichloroethane) as a solvent. Under the reaction conditions a side reaction between sodium hydroxide and EDC produces small but hazardous quantities of vinyl chloride:

$$C_2H_4Cl_2 + NaOH \rightarrow C_2H_3Cl + NaCl + H_2O$$

Ethylene   Vinyl
dichloride   chloride

An alternative nonreactive solvent has been identified which eliminates this hazard (Hendershot, 1987).

- Phase transfer catalysis processes (Starks and Liotta, 1978; Starks, 1987) for the synthesis of many organic materials use less, or sometimes no, organic solvent; may use less toxic solvent; may allow use of less hazardous raw materials (for example, aqueous HCl instead of anhydrous HCl); and may operate at milder conditions. Some types of reactions where phase transfer catalysis has been applied include:
—esterification
—nucleophilic aromatic substitution
—dehydrohalogenation
—oxidations
—alkylation
—aldol condensations

Rogers and Hallam (1991) provide other examples of chemical approaches to inherent safety, involving synthesis routes, reagents, catalysts and solvents.

Innovative chemical synthesis procedures have been proposed as offering potential for economical and environmentally friendly routes to a variety of chemicals. These novel chemical reactions also offer potential for increasing the inherent safety of processes by eliminating hazardous materials, eliminating chemical intermediates, or allowing less severe operating conditions. Some examples of interesting and potentially inherently safer chemistries include:

- Electrochemical techniques, proposed for the synthesis of naphthaquinone, anisaldehyde, and benzaldehyde (Walsh and Mills, 1993).
- Extremozymes—enzymes that can tolerate relatively harsh conditions, suggested as catalysts for complex organic synthesis of fine chemicals and pharmaceuticals (Govardhan and Margolin, 1995).
- Domino reactions, in which a series of carefully planned reactions occurs in a single vessel, used to prepare complex biologically active organic compounds (Hall, 1994; Tietze, 1995).
- Solid superacid catalysts, proposed as replacements for catalysts such as hydrogen fluoride and aluminum chloride for processes such as alkylation and acylation (Misono and Okuhara, 1993).
- Laser light "micromanaged" reactions, directed to the production of desired products (Flam, 1994).
- Supercritical processing, allowing the use of less hazardous solvents such as carbon dioxide or water in chemical reactions. This benefit must be balanced against the high temperatures and pressures required for handling supercritical fluids. Johnston (1994), DeSimone et al. (1994), and Savage et al. (1995) review some potential applications of supercritical processing.

Much of this chemistry is still at an early stage of research, and there are few, if any, commercial applications. However, the potential environmental and safety benefits of these and other innovative chemical synthesis techniques will encourage further research and development.

A United States Environmental Protection Agency report (Lin et al., 1994) contains an extensive review of inherently safer process chemistry options which have been discussed in the literature. This report includes chemistry options which have been investigated in the laboratory, as well as some which have advanced to pilot plant and even to production scale.

Solvents

Replacement of volatile organic solvents with aqueous systems or less hazardous organic materials improves safety of many processing operations and final products. In evaluating the hazards of a solvent, or any other process chemical, it is essential to consider the properties of the material at the processing conditions. For example, a combustible solvent is a major fire hazard if handled above its flash point or boiling point.

Some examples of solvent substitutions include:

- Water based paints and adhesives, replacing solvent based products
- Less volatile solvents with a higher flash point, used for agricultural formulations (Catanach and Hampton, 1992). In many cases, aqueous or dry flowable formulations for agricultural chemicals may be used instead of organic formulations
- Aqueous and semi-aqueous cleaning systems, used for printed circuit boards and other industrial degreasing operations (Mandich and Krulik, 1992; Davis et al., 1994)
- Abrasive media cleaning systems, replacing hazardous organic solvents for paint stripping (Davis et al., 1994)
- N-Methyl pyrrolidone, dibasic ethers, and organic esters, substituting for more hazardous paint removers (Paint Removers, 1991; Davis et al., 1994)

There has been an active effort to substitute inherently safer and more environmentally friendly solvents in many industries. Goldschmidt and Filskov (1990) and Sorensen and Peterson (1992) identify scores of solvent substitutions which have been made in a variety of industries, including food processing, textile, wood and furniture, printing, and casting. The United States Environmental Protection Agency is developing an expert system to aid in solvent substitution for the printing industry (Timberlake and Govind, 1994).

3.3. Moderate

Moderate means using materials under less hazardous conditions, also called attenuation. Moderation of conditions can be accomplished by strategies which are either physical (lower temperatures, dilution) or chemical (development of a reaction chemistry which operates at less severe conditions).

Dilution

Dilution reduces the hazards associated with the storage and use of a low boiling hazardous material in two ways—by reducing the storage pressure, and by reducing the initial atmospheric concentration if a release occurs. Materials which boil below normal ambient temperature are often stored in pressurized systems under their ambient temperature vapor pressure. The pressure in such a storage system can be lowered by diluting the material with a higher boiling solvent. This reduces the pressure difference between the storage system and the outside environment, reducing the rate of release in case of a leak in the system. If there is a loss of containment incident, the atmospheric concentration of the hazardous material at the spill location is reduced. The reduced atmospheric concentration at the source results in a smaller hazard zone downwind of the spill.

Some materials can be handled in a dilute form to reduce the risk of handling and storage:

- Aqueous ammonia or methylamine in place of the anhydrous material
- Muriatic acid in place of anhydrous HCl
- Dilute nitric acid or sulfuric acid in place of concentrated fuming nitric acid or oleum (SO_3 solution in sulfuric acid)

If a chemical process requires the concentrated form of the material, it may be feasible to store a more dilute form, and to concentrate the material by distillation or some other technique in the plant prior to introduction to the process. The inventory of material with greater intrinsic hazard (i.e., undiluted) is reduced to the minimum amount required to operate the process, but the distillation adds a new hazardous process.

Chemical reactions are sometimes conducted in a dilute solution to moderate reaction rates, to provide a heat sink for an exothermic reaction, or to limit maximum reaction temperature by "tempering" the reaction. In this example there are conflicting inherent safety goals—the solvent moderates the chemical reaction, but the dilute system will be significantly larger for a given production volume. Careful evaluation of all of the process risks is required to select the best overall system.

Refrigeration

Many hazardous materials, such as ammonia and chlorine, are stored at or below their atmospheric boiling points with refrigeration. Refrigerated storage reduces the magnitude of the consequences of a release from a hazardous material storage facility in three ways—by reducing the storage pressure, by reducing the immediate vaporization of leaking material and the subsequent evolution of vapors from the spilled pool of liquid, and by reducing or eliminating liquid aerosol formation from a leak.

Refrigeration, like dilution, reduces the vapor pressure of the material being stored, reducing the driving force (pressure differential) for a leak to the outside environment. If possible, the hazardous material should be cooled to or below its atmospheric pressure boiling point. At this temperature, the rate of flow of a liquid leak will depend only on liquid head or pressure, with no contribution from vapor pressure. The flow through any hole in the vapor space will be small and will be limited to breathing and diffusion.

Material stored at or below its atmospheric pressure boiling point has no superheat. Therefore there will be no initial flash of liquid to vapor in case of a leak. Vaporization will be controlled by the evaporation rate from the pool formed by the leak. This rate can be minimized by the design of the containment dike, for example, by minimizing the surface area of the liquid spilled into the dike area, or by using insulating concrete dike sides and floors. Because the spilled material is cold, vaporization from the pool will be further reduced.

Many materials, when released from storage in a liquefied state under pressure, form a jet containing an extremely fine liquid aerosol. The fine aerosol droplets formed may not rain out onto the ground, but instead may be carried downwind as a dense cloud. The amount of material contained in the cloud may be significantly higher than would be predicted based on an equilibrium flash calculation assuming that all of the liquid phase rains out. This phenomenon has been observed experimentally for many materials, including propane, ammonia, hydrogen fluoride, and monomethylamine. Refrigeration of a liquefied gas to a temperature near its atmospheric pressure boiling point eliminates the two-phase flashing jet, and the liquid released will rain out onto the ground. Containment and remediation measures such as spill collection, secondary containment, neutralization, and absorption may then be effective in preventing further vaporization of the spilled liquid (CCPS, 1993a).

Figure 3.4 is an example of a refrigerated storage facility for chlorine. This facility includes a covered spill collection sump which is covered to reduce evaporation to the atmosphere, both by containing the evapo-

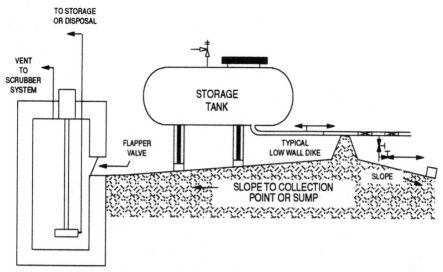

Figure 3.4. A chlorine storage system with collection sump with vapor containment (Puglionesi and Craig, 1991).

rating vapors and by reducing heat transfer from the surrounding atmosphere. The spill collection sump is vented to a scrubber which collects the chlorine vapor which evaporates from the sump.

Marshall et al. (1995) provide a series of case studies which evaluate the benefits of refrigerated storage for six materials—ammonia, butadiene, chlorine ethylene oxide, propylene oxide, and vinyl chloride. They conclude that "refrigerated storage is generally safer than pressurized storage" for all of the chemicals studied except ammonia. Ammonia was reported to be an exception "due to a density shift with temperature making it heavier than the surrounding air." Other materials may give similar results, and it is essential that the designer fully understand the consequences of potential incidents.

Less Severe Process Conditions

Processing under less severe conditions, close to ambient temperature and pressure, increases the inherent safety of a chemical process. Some examples include:

- Improvements in ammonia manufacturing processes have reduced operating pressures. In the 1930s ammonia plants operated at pressures as high as 600 bar. In the 1950s, process improvements had reduced operating pressures to 300–350 bar. By the 1980s, ammonia processes operating in the 100–150 bar range were being built. Besides being safer, the lower pressure plants are also cheaper and more efficient (Kharbanda and Stallworthy, 1988).

- Catalyst improvements allow methanol plants and plants using the Oxo process for aldehyde production to operate at lower pressures. The process also has a higher yield and produces a better quality product (Dale, 1987).

- Improvements in polyolefin manufacturing technology have resulted in lower operating pressures (Althaus and Mahalingam, 1992; Dale, 1987).

- Use of a higher boiling solvent may reduce the normal operating pressure of a process, and will also reduce the maximum pressure resulting from an uncontrolled or runaway reaction (Wilday, 1991).

- Semi-batch or gradual addition batch processes limit the supply of one or more reactants, and increase safety when compared to batch processes in which all reactants are included in the initial batch charges. For an exothermic reaction, the total energy of reaction available in the reactor at any time is minimized. However, the inherent safety benefits of semi-batch operation are only realized if the limiting reactant is actually consumed as it is fed, and there is no buildup of unreacted material. A number of process upsets, such as contamination with a reaction inhibitor, operating at too low a temperature, forgetting to charge a catalyst to the reactor, or forgetting to start the agitator, could result in buildup of unreacted material. If any of these upsets causing loss of reaction can occur, it is important to be able to ensure that the reactants are indeed being consumed as they are fed in order to realize the inherent safety benefits of a semi-batch process. The reactor could be monitored to provide confirmation that the limiting reactant is being consumed, by on-line analysis or by monitoring some physical property of the batch that is reliably correlated to reaction progress. (CCPS, 1993a)

- Advances in catalysis will result in the development of high yield, low waste manufacturing processes. Catalysts frequently allow the use of less reactive raw materials and intermediates, and less severe processing conditions. High yields and improved selectivity reduce the size of the reactor for a specified production volume. High selectivity for the desired product also reduces the size and complexity of the product purification equipment. It may be possible to develop a catalyst that is sufficiently selective that it becomes unnecessary to purify the product at all, as in a process for HCFC-141b (CH_3CFCl_2) described by Manzer (1993). Allen (1992), Manzer (1993, 1994), and Dartt and Davis (1994) describe a number of catalytic processes which are potentially environmentally friendly and safer.

Secondary Containment—Dikes and Containment Buildings

Secondary containment systems are best described as passive protective systems. They do not eliminate or prevent a spill or leak, but they can significantly moderate the impact without the need for any active device. Also, containment systems can be defeated by manual or active design features. For example, a dike may have a drain valve to remove rain water, and the valve could be left open. A door in a containment building could be left open.

Harris (1987) provides an excellent set of guidelines for the design of storage facilities for liquefied gases to minimize the potential for vapor clouds:

- Minimize substrate surface wetted area.
- Minimize pool surface open to atmosphere.
- Reduce heat capacity and/or thermal conductivity of substrate.
- Prevent "slosh over" of containment walls and dikes.
- Avoid rainwater accumulation.
- Keep liquid spills out of sewers.
- Shield the pool surface from the wind.
- Provide vapor removal system to a scrubber or other emission control device.
- Provide liquid recovery system to storage where possible.
- Avoid direct sunshine on containment surfaces in hot climates.
- Direct spills of flammable materials away from pressurized storage vessels to reduce the risk of a boiling liquid expanding vapor explosion (BLEVE).

Figure 3.5 shows a liquefied gas storage facility which incorporates many of these principles. CCPS (1993a) provides several other exam-

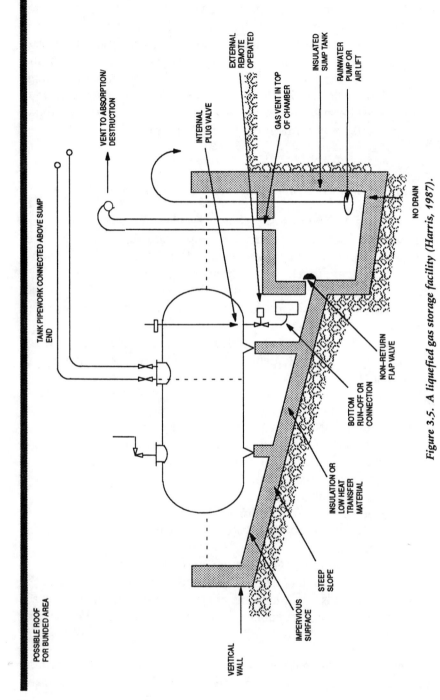

Figure 3.5. A liquefied gas storage facility (Harris, 1987).

ples for storage facilities, including chlorine, flammable liquids and liquefied flammable gases.

Containment buildings have been used to limit the impact of loss of containment incidents for many toxic materials, including chlorine and phosgene (CCPS, 1993a). Containment buildings can cover a wide range of structures, from a simple, light structure to reduce evaporation of a spill of a relatively nonvolatile toxic material, to a very strong pressure vessel designed to withstand an internal explosion. Englund (1991a) describes the evolution in the design of a phosgene handling facility from an open air plant through various stages of increasing containment, culminating in the design of Figure 3.6. The process is totally enclosed in a large pressure vessel capable of withstanding the overpressure in case of a flammable vapor deflagration inside the containment vessel.

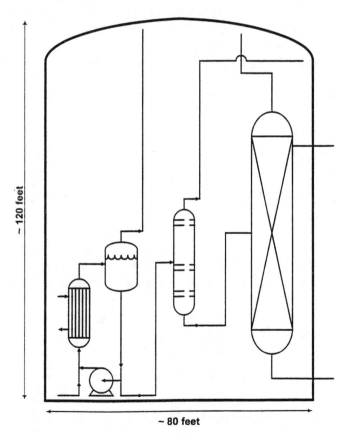

Figure 3.6. A chemical process totally contained in a large pressure vessel (based on Englund, 1991a)

Containment buildings are an example of inherent safety conflicts and tradeoffs. A containment building provides protection outside the building, but it can also trap and concentrate material from small leaks inside the building, increasing the risk to personnel entering the building.

Provisions must be made to ensure worker protection for a process located in a containment building. For example, the atmosphere in the containment structure should be monitored for hazardous vapors, operations should be remotely controlled from outside the containment structures, access should be restricted, and proper personal protective equipment should be used when entry into the containment structure becomes necessary.

In particular, great care must be take when evaluating tradeoffs for a containment building for a flammable and toxic material such as hydrogen cyanide. A leak or fire inside the building could cause a confined vapor cloud explosion which destroys the building. The total risk may actually increase.

Frank (1995), and Purdy and Wasilewski (1995) have published quantitative risk studies which evaluate the benefits of containment buildings for chlorine handling facilities.

3.4 Simplify

Simplify means designing to eliminate unnecessary complexity, reducing the opportunities for error and misoperation. A simpler plant is generally safer and more cost effective than a complex one. For example, it is often cheaper to spend a relatively small amount of money to build a higher pressure reactor, rather than a large amount of money for an elaborate system to collect and treat the discharge from the emergency relief system of a reactor designed for a lower maximum pressure. A few examples of simplification and error tolerance are discussed in the following sections. Kletz (1991b) provides additional examples. Others are found in Chapter 4 of this book.

> **In all cases where we are free to choose between easy and difficult modes of operation, it is most rational to choose the easy.**
>
> —Thomas Jefferson, 1784

Containment within Process Equipment

In many cases it is possible to design process equipment strong enough to contain the maximum or minimum pressure resulting from a process incident (CCPS 1993a). Containment within the process vessel simplifies the design by eliminating high pressure interlock systems. Emergency relief devices such as rupture disks or relief valves may still be required by regulations and codes, but the size may be reduced and the hazards associated with opening of the relief devices may be considered to be eliminated. Catch tanks, scrubbers, flare stacks, or other devices to dispose of the effluent from emergency relief systems safely may also be eliminated.

Combustion

The maximum pressure resulting from a deflagration of a combustible dust or flammable vapor in air initially at atmospheric pressure is often less than 10 bar. It may be feasible to build equipment strong enough to contain this type of event. When designing a system for combustion containment, the engineer must consider factors such as highly reactive materials, oxygen or other oxidant enriched atmospheres, and congested geometry inside vessels or pipelines which could result in transition to detonation. All of these factors can significantly increase the maximum pressure of a combustion reaction.

Vacuum

Designing vessels for full vacuum eliminates the risk of vessel collapse due to vacuum. Many storage and transport vessels have been imploded by pumping material out with the vents closed.

Runaway Reactions

Choosing a reactor design pressure sufficiently high to contain the maximum pressure resulting from a runaway reaction eliminates the need for a large emergency relief system. It is essential that the reaction mechanisms, thermodynamics, and kinetics under runaway conditions are thoroughly understood for the designer to be confident that the design pressure is sufficiently high for all credible reaction scenarios. All causes of a runaway reaction must be understood, and any side reactions, decompositions, and shifts in reaction paths at the elevated temperatures and pressures experienced under runaway conditions must be evaluated. Many laboratory test devices and procedures are available for evaluating the consequences of runaway reactions (CCPS, 1995d, e).

Containment Vessels

In many cases, if it is not feasible to contain a runaway reaction within the reactor, it may be possible to pipe the emergency device effluent to a separate pressure vessel for containment and subsequent treatment.

Heat Exchangers

The shell and tube sides of heat exchangers can be designed to contain the maximum attainable pressure on either side, eliminating reliance on pressure relief to protect the exchanger shell in case of tube rupture.

Liquid Transfer

Liquid transfer systems can be designed to minimize leakage potential. For example, transfer systems which use gravity, pressure, or vacuum require no moving parts or seals. If a pump is needed, centrifugal pumps with double mechanical seals, diaphragm pumps, jet pumps, and various types of sealless pumps may be good choices. Sealless pumps greatly reduce the risk of a process fluid leak, but they also introduce new hazards and concerns, such as overheating, which may be very rapid, and internal leakage.

Reactor Geometry

Research on safer nuclear power reactors has identified systems which utilize natural convection to provide emergency core cooling rather than relying on pumped cooling water circulation. Other new approaches utilizing reactor geometry, in-situ moderators, and novel materials of construction can prevent core overheating more reliably and are being researched (Forsberg et al., 1989).

Similar approaches are applicable in the chemical industry. For example, maleic anhydride is manufactured by partial oxidation of benzene in a fixed catalyst bed tubular reactor. There is a potential for extremely high temperatures due to thermal runaway if feed ratios are not maintained within safe limits. Catalyst geometry, heat capacity, and partial catalyst deactivation have been used to create a self-regulatory mechanism to prevent excessive temperature (Raghaven, 1992).

Fail Safe Valves

Processes should be reviewed to identify the safest failure position for all electric or pneumatic valves. The designer should consider all failures including the control system, all driving utilities, and all operating situations. In most cases process valves should fail closed. Often cooling water valves should fail open. In some cases a valve should fail in its last position (in-place), remaining open if it is already open and remaining closed if it is already closed. For example, the vent valve on a batch reactor which is vented to a scrubber in several steps, but must be closed for a pressurized reaction step, should probably fail in its last position.

Remember that the failure position of a valve refers to its failure mode if there is a utility failure. A valve can mechanically fail in any position; it is possible for a "fail closed" valve to get stuck in the open position. When doing a process hazard analysis it is important to consider all possible failure positions of a valve, and not only the failure position resulting from utility failure.

Distributed Control Systems

A distributed control system (DCS) normally uses input and output modules which contain eight, sixteen, or more inputs or outputs. Failure of the module will simultaneously disable a large number of control loops. Attention to the assignment of input/output points to the modules makes the plant more tolerant of a failure of an input or output module (CCPS, 1993a). For a more detailed discussion of process control systems, see the process control part of Section 4.4, and Sections 6.4 and 6.5.

Separation of Process Steps

A multistep batch process can be carried out in a single vessel, or in several vessels, each optimized for a single processing step. The complexity of the batch reactor in Figure 3.7, with many potential process fluid and utility interactions, can be greatly reduced by dividing the same process into three vessels as shown in Figure 3.8. Again, this is an example of an inherent safety conflict. The system in Figure 3.7 requires only one reactor, although it is extremely complex, and process intermediates never leave the reaction vessel. The system in Figure 3.8 uses three vessels, each of which can be optimally designed for a single task. Although each vessel is considerably simpler, it is necessary to transfer intermediate products from one vessel to another. If one of those intermediate products is extremely toxic, it may be judged to be

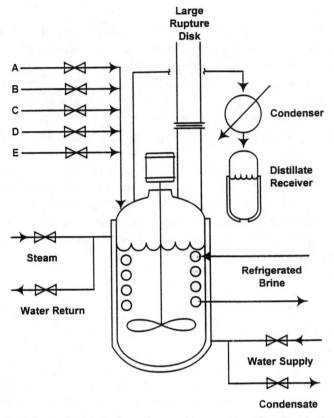

Figure 3.7. *A complex batch reactor for a multistep process (from Hendershot, 1987).*

preferable to use the single reactor (a "one pot" process) to avoid transfer of the toxic intermediate. As always, the inherent safety advantages and disadvantages of each system must be evaluated with careful consideration of all of the hazards of a particular chemical process, as discussed in Section 2.5.

3.5. Summary

This chapter describes the four main design strategies for development of inherently safer processes:

- Minimize
- Substitute
- Moderate
- Simplify

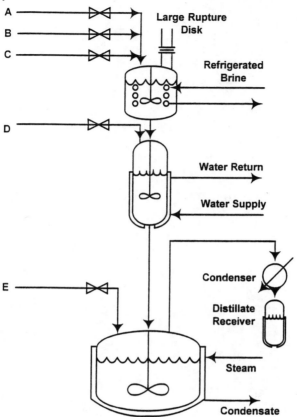

Figure 3.8. The same process as Figure 3.7 in a series of simpler reactors (from Hendershot, 1987).

These strategies can be applied at any phase of the process life cycle. Examples of each strategy are given, drawn from all phases in the life cycle. In the next chapter, we will shift our focus and discuss opportunities for application of the strategies described at specific stages in the overall life cycle of a chemical process.

4

Life Cycle Stages

4.1 General Principles across All Life Cycle Stages

A process goes through various stages of evolution; research, process development, design and construction, operations, maintenance, modifications, and, finally, decommissioning. Progression through these stages is typically referred to as the process life cycle and is shown in Figure 4.1. Throughout a process's life cycle stages the opportunity to apply continuously the philosophies and practices of inherently safer technologies and strategies must be recognized and cultivated. The pressures of social opinion and regulatory demands have historically been a result of negative experiences in the industry. As the chemical process industry evolves, its successes and expectancy will be judged by our ability to incorporate the best technologies to guarantee the protection of employees, the community, and the environment.

This chapter demonstrates that enhanced process safety, with economic and other holistic objectives—for example, quality, productivity, energy conservation, and pollution prevention—can be achieved by applying inherently safer philosophies and strategies. This application, utilizing formal review methods by trained individuals as identified in Chapter 7, will link the general principles of inherently safer concepts to all life cycle stages. Beginning the review process at the earliest stages of research and recognizing that enhancements can be

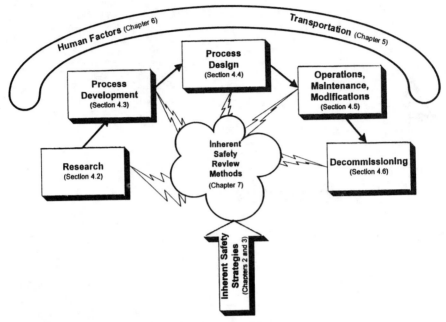

Figure 4.1.   The process life cycle stages.

achieved during all stages of the life cycle are necessary to assure that the highest level of integrity is maintained. This approach with an awareness for transportation issues and human factors is clearly necessary for the chemical industry of the future to prosper.

4.2. Research

Research chemists working with other R&D professionals play a fundamental role in applying inherent safety in the development of a process. They are responsible for decreasing to a practical minimum the risks imposed by a future product and by the process used to make and distribute a product. Importantly, they have this responsibility early in the development of a process and can significantly affect the outcome. Kletz (circa 1988) has pointed out that, relatively speaking, a safety problem eliminated by the use of inherent safety will cost $1 to fix at the research stage, $10 at the process flow sheet stage, $100 at the final design stage, $1000 at the production stage, and $10,000 at the postincident stage. Thus, the early application of inherent safety pays dividends.

> **An ounce of Inherent Safety is worth a pound of life cycle cost.**

To apply inherent safety appropriately, research chemists must make an in-depth investigation into the process chemistry and into the entire process that may develop based on that chemistry. An adequate investigation necessitates input from a diverse team of people, including research chemists and business, engineering, safety, environmental personnel. They must consider the impact that the use of a particular process chemistry will have on a wide range of populations. These include the ultimate customer of the product, process operating personnel, the general public, and potentially impacted plant and animal populations. To chose the "inherently safest" chemistry, the team needs to take into account:

- the environment
- process hazards
- worker safety
- upstream, downstream, and ancillary unit operations (including waste disposal)
- inventory needs
- transportation of raw materials and final products

Research management has the responsibility to create a working environment that promotes, encourages, rewards, and facilitates the application of inherent safety in the development of process chemistries.

Inherently Safer Synthesis

Research chemists have many opportunities to incorporate inherent safety in the choice of chemical synthesis route, including:

- Catalysis, leading to less severe operating conditions, the use of a less reactive reagent, or the opportunity to eliminate or reduce the use of a hazardous solvent.
- Enzyme-based chemistry and biosynthesis.
- Immobilization of hazardous reagents and catalysts by attaching active groups to polymeric or immobile backbones.
- Reduction of reaction hazards by tempering a reaction with the use of a more volatile solvent that will boil and more reliably remove the heat of reaction.
- Reagents that are diluted.

- Reactions in water as opposed to a those which proceed in a hazardous solvent.
- Elimination of hazardous unit operations (see Section 4.3).
- Supercritical processing (see Section 3.2).
- Replacement of batch reaction processes with semi-batch or continuous processes reducing the quantity of reactant present.
- Use of processes that are less sensitive to critical operating parameter variations.

The wide array of choices available to research chemists necessitates a diligent search for hazards to select the inherently safest chemistry. One of the means to search for hazards is to conduct a literature search, looking in particular for reports of incidents occurring in processes using the same or similar process chemistry being considered.

If it has blown up before, look for an alternative.

Types of Hazards and Hazardous Events

Table 4.1 is a representative list of the types of hazards and hazardous events that research chemists are attempting to address in searching for the best chemistry. Some key factors to consider relative to process hazards include:

- Important flammability characteristics are the lower and upper flammability limits, the flash point, the minimum ignition energy, the minimum oxygen concentration, and the autoignition temperature. Values of some of these properties are published for many compounds (NFPA, 1994) . These numbers have typically been developed under standardized test conditions. Process conditions may influence their values.
- The fireball resulting from ignition of a cloud of flammable vapor may be relatively long lasting (2–5 seconds), and represents a thermal radiation hazard to those close to the cloud CCPS (1994b).
- Vapor cloud explosions can cause damaging overpressures (CCPS, 1994b).

TABLE 4.1
Types of Hazards and Hazardous Events

FIRES

    Flash fires
    Pool (large, sustained) fires
    Jet fires

EXPLOSIONS

    Vapor clouds

    Confined deflagrations

    Detonations

    Pressure Vessel Ruptures:
        Exothermic runaway reactions
        Physical overpressure of pressure vessels
        Brittle fracture
        Polymerizations
        Decompositions
        Undesired reactions catalyzed by materials of construction or by ancillary
            materials such as pipe dope and lubricants
        Boiling liquid, expanding vapor explosions (BLEVEs)

TOXICITY RELATED HAZARDS

    Environmentally toxic to plant, animal or fish:
        Chronic or acute
        Toxic to individual species or broadly hazardous
        Pesticides, fungicides, herbicides, insecticides, fumigants

    Toxic to humans
        Chronic or acute
        Reversible injury or irreversible injury or death
        Carcinogens
        Endocrine modifiers (e.g., estrogen mimics)
        Persistent bioaccumulative toxins (PBTs)

    Long-term environmental hazards:
        Greenhouse gases
        Ozone depletors

PRODUCT HAZARDS

    Customer injury
    Waste disposal environmental hazard

- A flammable vapor explosion in a vessel initially at atmospheric pressure can create a pressure of 10 atmospheres (NFPA 69, 1992, Section 5-3.3.1). A weak process vessel requires substantial vent area for vapor–air explosion relief.
- A flame in a pipeline containing a flammable mixture can transition to a detonation. This flame-to-detonation transition is discussed in Lewis and von Elbe (1987, Section 8.5).

- Pool fires are of long duration and the radiation intensity near the pool is high. Storage vessels exposed to a pool fire may explode. Exposure of a storage vessel to a fire is one cause of a BLEVE—boiling liquid expanding vapor explosion. This can occur if the pressure above a mass of liquid at a high temperature and high vapor pressure suddenly drops to ambient by the catastrophic failure of the storage vessel (CCPS, 1994b). The shock wave produced by the flashing of the superheated liquid is destructive. If the liquid is flammable, there will likely be a subsequent fireball and an unconfined vapor cloud explosion.
- An unvented runaway reaction in a vessel or a physical over-pressurization of a vessel can cause it to lose its structural integrity. There are numerous methods to calculate the energy of the explosion of such a vessel. The reaction stability is a complex function of temperature, concentration, impurities, and degree of confinement. Knowledge of the reaction onset temperature, the rate of reaction as a function of temperature, and heat of reaction is necessary for analysis of a runaway reaction. Minimum exothermic runaway temperature and runaway reaction consequences can be characterized by thermal stability screening. CCPS (1995e) presents additional information on testing techniques and apparatus for the reactivity of chemicals.
- Toxicological effects are expressed in terms of the population affected, whether the effect is acute or chronic, the route of entry, dosage, and extent and type of injury. A discussion of toxicological effects is found in Patty's (1993, Vol. II). Rand and Petrocelli (1985) provide an overview of the types of compounds that are harmful to the aquatic environment and the manner by which they are harmful.
- Parshall (1989) notes that product liability issues are complex and varied. An attorney knowledgeable in product liability issues may be an important contributor to process hazards discussions if such issues may affect the advisability of a proposed process chemistry.

Tools for Hazards Identification

The identification of hazards includes both a search for those hazards reduced or eliminated by inherently safer design, and a search for hazards controlled by instrumentation and administrative procedures.

Research chemists cannot do these searches independently. There are a number of tools designed to identify and evaluate hazards. Several of these "identification tools" are described below.

Molecular Structure and Compounds

Certain molecular groupings are likely to introduce hazards into a process. The research chemist should identify groupings and molecular structures that may introduce these hazards. A search of the open literature will assist in identifying which types of compounds are likely to create potential hazards. Table 4.2 presents molecular structures and compound groupings associated with known hazards. The groupings in the table were developed from CCPS (1995d, Table 2.5), and Medard (1989). The table is not all-inclusive.

The hazards of new compounds may not be known but the hazards of analogous compounds, or of those with the same or similar molecular groupings may be known. Testing may be necessary to determine hazardous characteristics.

Reactivity of Types of Compounds

Many additional hazards result from the hazardous reactivity of combinations of chemicals. The open literature contains numerous lists of the reactivity of different types of chemical combinations. Table 4.3 presents examples of combinations of compounds which are known to be reactive. More complete discussions and lists of highly energetic chemical interactions are found in CCPS (1995d, especially Table 2.14), Yoshida (1987), Medard (1989), FEMA (Appendix D, ca. 1989), and Bretherick (1995).

The Interaction Matrix

The chemical interaction or reaction matrix is a recognized useful hazard identification tool. Figure 4.2 on page 64 is a conceptual matrix. Matrices are referenced in CCPS (1992, pp. 45–47 and 242–245). An example matrix for a process is presented in Gay and Leggett (1993). Necessary utilities, such as inerting nitrogen, and the materials of construction should be listed as components in a matrix, as should the operator and other populations impacted by the process. Tests or calculations may be appropriate if the effect of an interaction identified in a matrix is unknown. The CHETAH software program developed by ASTM (ASTM, 1994) may also assist in this determination. A complete review of the topic of reactivity is also found in CCPS (1995d,e).

TABLE 4.2

Representative Potentially Hazardous Molecular Groupings

The primary associated hazard is indicated to the right of the molecular grouping.	
Ammonia:	Toxicity and fire
Chlorinated hydrocarbons:	Toxicity
Cyano compounds:	Toxicity
Double and triple bonded hydrocarbons:	Fire and explosion
Epoxides:	Explosion
Hydrides and hydrogen:	Explosion
Metal acetylides:	Explosion

Nitrogen compounds:
    Amides and imides and nitrides
    Azides
    Azo- and diazo- and diazeno- compounds
    Difluoro amino compounds
    Halogen-nitrogen bond containing compounds
    Hydrazine-derived nitrogen compounds
    Hydroxy ammonium salts
    Metal exolates
    Nitrates (including ammonium nitrate)
    Nitrites
    Nitroso compounds
    N-metal derivatives
    Polynitro alkyl and aryl compounds:
    Sulfur–nitrogen bond containing compounds
	All explosion

Oxygenated compounds of halogens:	Explosion
Oxygenated manganese compounds:	Explosion
Peroxides (and peroxidizable compounds):	Fire and explosion
Polychlorinated biphenyls (PCBs):	Environmental
Polycyclic aromatic hydrocarbons (PAH's):	Environmental

The preparation of a matrix and the subsequent evaluation of the hazards identified can lead to a qualitative judgment of process risk and to the identification of available pathways to reduce that risk. Software is available to assist in making and maintaining interaction-like matrices. One example is a database shell called CHEMPAT (AIChE, 1995). When CHEMPAT is customized by the user, a compatibility chart is produced based on user-supplied chemical information.

TABLE 4.3
Reactive Combinations of Chemicals

Substances				Type of Hazard
A	**+**	**B**	**→**	**Hazardous Event**
Acids		Chlorates		Spontaneous ignition
		Chlorite and Hypochlorite		Spontaneous ignition
		Cyanides		Toxic and flammable gas generation
		Fluorides		Toxic gas generation
		Epoxides		Heat generation, polymerization
Combustibles		Oxidizing Agents		Explosion
		Anhydrous Chromic Acid		Spontaneous ignition
		Potassium Permanganate		Spontaneous ignition
		Sodium Peroxide		Spontaneous ignition
Alkali		Nitro compounds		Easy to ignite
		Nitroso Compounds		Easy to ignite
Ammonium Salts		Chlorates		Explosive ammonium salts formed
		Nitrites		Explosive ammonium salts formed
Alkali Metals		Alcohols, Glycols		Flammable gas and heat generation*
		Amides, Amines		Flammable gas and heat generation*
		Azo- and diazo- compounds		Flammable gas and heat generation*
Inorganic Sulfide Metals		Water		Toxic and flammable gas generation
		Explosives		Heat generation and explosion
		Polymerizable Compounds		Polymerization and heat generation

*Spontaneous ignitions also possible (e.g. MEOH plus K)

Other Hazards Identification Tools
The "What if . . ." method, the checklist, and HAZOP are well-publicized hazard identification tools. CCPS (1992) presents guidance on the use of these tools.

The use of any of the above techniques demands knowledge, experience, and flexibility. No prescriptive set of questions or key words or list is sufficient to cover all processes, hazards, and all impacted populations. As a research chemist reviews a chemistry and its potential application, there are advantages to maintaining an open mind when applying the various techniques designed to open up avenues of thought. The reader is referred to Chapter 7 for additional information and direction on the choice of process hazard review techniques.

	chemicals	materials of construction	utilities	lifeforms
chemicals	✲	✲	✲	✲
materials of construction		✲	✲	✲
utilities			✲	✲
lifeforms				✲

NOTES:

- Chemicals include all raw materials, intermediate, product, and by-product chemicals, as well as any other chemicals used in the process (for example, catalysts).
- Materials of construction include ancillary materials like pipe dope.
- Utilities include ambient air.
- Lifeforms include the operator, the plant neighbor, the general public, the ultimate consumer of the product. They also include the aquatic and terrestrial plant, fish and animal life.
- The ✲ represents a reference to notes. These notes should be sufficiently complete to highlight the type of hazard and the degree (extent, severity) of the hazard.

Figure 4.2. A conceptual interaction matrix.

Alternatives

Alternative process chemistries will likely pose different hazards and different degrees of hazard. The comparison of alternatives will inevitably lead to "tradeoffs." The chosen candidate, although not devoid of hazards altogether, will likely be the one with the least risk. There is at present no universally accepted "tool" to judge among alternatives (see discussion in Sections 2.4 and 2.5). Edwards et al. (1996) propose such a tool. The result is a "score" for each alternative, with the alternative with the lowest score being the most inherently safe. Also, the Dow Fire and Explosion Index (FEI) (Dow, 1994b), the Dow Chemical Exposure Index (Dow, 1994a), the Mond Index (Lewis, 1979; ICI, 1985), and the Mond-like acute Toxicity Hazard Index (Tyler et al., 1996) are measures that can assist in choosing the most appropriate process chemistry. Guidance on alternative selection is recognized as a future need in Chapter 8.

The Chemist and Real Business . . . Life Cycle Cost

The life cycle cost of a process is the net total of all expenses incurred over the entire lifetime of a process. The choice of process chemistry can dramatically affect this life cycle cost. A quantitative life cycle cost cannot be estimated with sufficient accuracy to be of practical value. There is benefit, however, in making a qualitative estimate of the life cycle costs of competing chemistries. Implicit in any estimate of life cycle cost is the estimate of risk. One alternative may seem more attractive than another until the risks associated with product liability issues, environmental concerns, and process hazards are given due consideration. Value of life concepts and cost–benefit analyses (CCPS, 1995a, pp. 23–27 and Chapter 8) are useful in predicting and comparing the life cycle costs of alternatives.

Examples of Synthesis Routes Inherently Safer Than Others

As summarized by Bodor (1995), the ethyl ester of DDT is highly effective as a pesticide and is not as toxic. The ester is hydrolytically sensitive and metabolizes to nontoxic products. The deliberate introduction of a structure into the molecule which facilitates hydrolytic deactivation of the molecule to a safer form can be a key to creating a chemical product with the desired pesticide effects but without the undesired environmental effects. This technique is being used extensively in the pharmaceutical industry. It is applicable to other chemical industries as well.

Synthetic rubber latex was made by a process with a large and hazardous inventory of butadiene and styrene. In a modified process, the reactor has an initial charge of water and emulsifier. Also, the monomers are added to the reactor as one premixed stream and the emulsified aqueous sodium persulfate is added as the other stream. The improved scheme, discussed by Englund (1991a) contains less hazardous material and at a lower, more controllable temperature. It illustrates that large and established processes may be made safer by applying inherent safety.

4.3. Process Development

Process chemists and process engineers have essential roles to play in applying inherently safer strategies during the process development stage. The chemistry has already been established, thus defining the hazards of the materials. Process development personnel need to focus primarily on process synthesis, unit operations, and the type of equipment required for an inherently safer process. A thorough understanding of the necessary operational steps and alternate operational steps is essential to develop an efficient and safe process.

In the process development stage we must realize the importance of creative thinking by continuing to look for better solutions and not stopping at the first solution identified. The old argument of "this is the way it has been done for years" is not sufficient reason to ignore the possibility that there might be better and safer processes for the future. The penalty for standing still is that your competitors might be advancing by developing the better technology. The process development stage is an opportunity to improve beyond the competitors' technology. This improvement will most likely require additional early work "up front" in considering and evaluating alternatives, but this effort will yield returns such as a safer and more cost effective final design.

Creative thinking is important not only in new process development, but also in continually reviewing and reevaluating existing processes for opportunities to make the process inherently safer. Many of the tools and techniques discussed in Section 4.2, on Research, are useful in the process development stage as well. It is appropriate to revisit the basic chemistry to study alternate options.

Risk Ranking Tools

Tools are available to assist in comparing the risk associated with two or more different processes. For example, the first sheet of the Dow Fire and Explosion Index (FEI) (Dow, 1994b) ranks the safety characteristics of the process from a fire/explosion standpoint, without taking credit for protective and mitigation features. The Dow Chemical Exposure Index (CEI) (Dow,1994a) and ICI's Mond Index (ICI, 1985; Tyler, 1985) are other ranking tools.

A number of vendors offer software based hazard assessment tools that help determine the magnitude of the hazards involved. With this software, calculations can be made to reflect the hazard for various failures. Some risk ranking software combines hazard assessment with probabilities of occurrence so that the relative risk levels can be assessed.

Unit Operations—General

There are a variety of ways of accomplishing a particular unit operation. Alternative types of process equipment have different inherently safer characteristics such as inventory, operating conditions, operating techniques, mechanical complexity, and forgiveness (i.e., the process/unit operation is inclined to move itself toward a safe region, rather than unsafe). For example, to complete a reaction step, the designer could select a continuous stirred tank reactor (CSTR), a small tubular reactor, or a distillation tower to process the reaction.

Before studying alternative types of equipment, we need to understand the critical process requirements. Is a solvent necessary? Must products or by-products be removed to complete the reaction? What mixing and/or time requirements are necessary? Once again, early work "up front" is needed before alternate reaction schemes can be evaluated.

Similarly, different unit operations are available to accomplish the same processing objective. For example, a filter, a centrifuge, or a decanter could be used to separate a solid from a liquid. Crystallization or distillation could also be used for purification.

It is inherently safer to develop processes with wide safe operating limits that are less sensitive to variations in critical safety operating parameters, as shown in Figure 4.3. Sometimes this type of process is referred to as a "forgiving" or "robust" process. If a process must be controlled within a very small temperature band in order to avoid

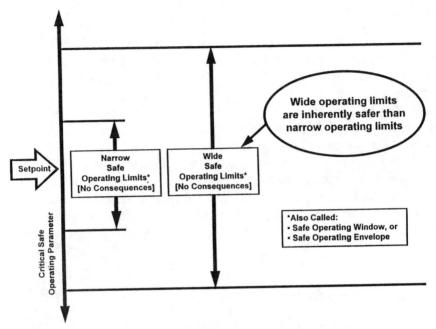

Figure 4.3. Process design safe operating limits.

hazardous conditions, that process would have narrow safe operating limits; a process with a larger temperature band will have wider safe operating limits. For some reactions, using an excess of one reactant can enlarge the safe operating limits. Determination of the size of the safe operating limits requires a complete understanding of the processes under consideration. Other synonyms for "safe operating limits" are "safe operating envelope" or "safe operating window."

Unit Operations—Specific

Some examples and considerations for specific common unit operations follow.

Reactors are where the action should be.

Reaction
Reactor design is particularly critical, because reactors involve chemical transformations and often potentially significant energy releases.

Evaluation of the safety characteristics for a given reactor design requires an understanding of what controls the rate of reaction (catalysis, mass transfer, heat transfer, etc.) as well as the total potential energy involved in the reaction. The possibility of the energy generating pressure and/or undesired side reactions should also be evaluated. This information is usually necessary for evaluating the suitability of various reactor types (CSTR, batch, tubular, various novel designs such as eductors in loop reactors, static mixers, extruders) for the desired reaction. Mixing and mass transfer are often the critical elements in reactor design as chemicals often react quickly once the molecules are brought together.

Not all reactions take place in a designated reactor. Some occur in a heat exchanger, a distillation column, or a tank. Understand the reaction mechanisms and know where the reactions occur before selecting the final design.

Be sure you understand where the reaction occurs.

Some batch reactions have the potential for very high energy levels. If all the reactants (and sometimes catalysts) are put into a kettle before the reaction is initiated, some exothermic reactions may result in a "runaway." The use of continuous or "semi-batch" reactors to limit the energy present and to reduce the risk of a "runaway" should be considered. The term "semi-batch" refers to a system where one reactant and, if necessary, a catalyst is initially charged to a batch reactor. A second reactant is subsequently fed to the reactor under conditions such that an upset in reacting conditions can be detected and the flow of the reactant stopped, thus limiting the total amount of potential energy in the reactor.

Additional discussion regarding reactor design strategies is covered in Section 3.1 on minimization (as an inherently safer design strategy), and in Section 4.4 on Design and Construction.

Distillation
There are options to minimize the hazards when distilling materials that may be thermally unstable or have a tendency to react with other chemicals present. Some options include:

- Trays without outlet weirs
- Proprietary designs and sieve trays

- Wiped film evaporators
- An internal baffle in the base section to minimize hold-up
- Reduced base diameter (Kletz, 1991b)
- Vacuum distillation to lower temperatures
- Smaller reflux accumulators and reboilers (Dale, 1987)
- Internal reflux condensers and reboilers where practical (Dale, 1987)
- Column internals that minimize holdup without sacrificing operation efficiency (Dale, 1987).

Another option is to remove toxic, corrosive, or otherwise hazardous materials early in a distillation sequence, reducing the spread of such materials throughout a process (Wells and Rose, 1986).

Low-inventory distillation equipment, such as the thin film evaporator, are also available and should be considered for hazardous materials. This equipment offers the additional advantage of short residence time and is particularly useful for reactive or unstable materials.

Solids Handling

Solids handling frequently has the potential for dusting, which can lead to potential health and explosion hazards. Handling solids in the form of larger particle size granules or pellets rather than a fine powder reduces the potential for worker exposure. Worker exposure hazards are reduced by formulating dyes as liquids or wet pastes rather than dry solids or powders (Burch, 1986).

If the solid is combustible, the dust explosion hazard can be greatly reduced or even eliminated by using a larger particle size material. It is important to remember, however, that particle attrition can occur during handling and processing, resulting in the generation of small particles which could increase dust explosion hazards. Study the sequence of size reduction steps or even the required particle size to minimize the number of processing steps that involve very small particles. Another option would be to change the form to a less dusty shape (pellets, beads, prills, etc.). Handling of solids as a wet paste or slurry can also reduce hazards. For example, using wet benzoyl peroxide instead of dry reduces the hazards of this extremely reactive material (Yoshida et al., 1991).

It may even be possible to eliminate solids handling by processing in a solution. However, this may require an assessment of the hazards of a toxic or flammable solvent in a process compared to the hazards of the solvent-free process.

Inherently safer approaches to dust explosion hazard control include inerting and building equipment strong enough to contain an explosion.

Heat Transfer

Some processes have large heat transfer requirements. This may result in large inventories of material within the heat transfer equipment. If the material is thermally unstable it would be inherently safer to reduce the residence time in the heat exchanger. Options to minimize heat exchanger inventory include the use of different types of heat exchangers. Inventories in shell and tube heat exchangers can be reduced by the use of "turbulators" in the tubes to enhance heat transfer coefficients, and by placing the more hazardous material on the tube side.

Heat transfer equipment has a great variation in heat transfer area per unit of material volume. Table 4.4 compares the surface compactness of a variety of heat exchanger types. Falling film evaporators and wiped film heat exchangers also reduce the inventory of material on the tube side. Process inventory can be minimized by using heat exchangers with the minimum volume of hazardous process fluid for the heat transfer area required.

Transfer Piping

Inventory in transfer lines can be a major risk. For example, a quantitative risk analysis of a chlorine storage and supply system

TABLE 4.4
Surface Compactness of Heat Exchangers
(Adapted from Kletz, 1991b)

Type of Exchanger	Surface Compactness (m^2/m^3)
Shell and tube	70–500
Plate	120–225 up to 1,000
Spiral plate	Up to 185
Shell and finned tube	65–270 up to 3,300
Plate fin	150–450 up to 5,900
Printed circuit	1,000–5,000
Regenerative—rotary	Up to 6,600
Regenerative—fixed	Up to 15,000*
Human lung	20,000

*Some types have a compactness as low as 25 m^2/m^3.

identified the pipeline from the storage area to the manufacturing area as the most important contributor to total risk (Hendershot, 1991b). To minimize the risk associated with transfer lines, their length should be minimized by careful attention to unit location and pipe routing. Pipe size should be sufficient to convey the required amount of material and no larger. However, small bore piping is less robust and less tolerant of abuse when compared to large diameter piping, and additional attention to proper support and installation will be required. In some cases, for example, chlorine for water treatment applications, it may be possible to transfer material as a gas rather than a liquid with a large reduction of inventory in the transfer line.

Piping systems should be designed to minimize the use of components that are likely to leak or fail. Sight glasses and flexible connectors such as hoses and bellows should be eliminated wherever possible. Where these devices must be used, they must be specified in detail so they are structurally robust, compatible with process fluids, and installed to minimize the risk of external damage or impact.

Where flanges are necessary, spiral wound gaskets and flexible graphite type gaskets are preferred. The construction of these gaskets makes them less likely to fail catastrophically resulting in a large leak. Proper installation of spiral wound gaskets, particularly torquing of the flanges, is important in preventing leaks.

4.4. Design and Construction

As the process moves from the process development stage to the design and construction stage the chemistry, unit operations, and type of equipment have been set. The design and construction stage needs to focus primarily on equipment specifications, piping and instrumentation design, installation details, and layout for an inherently safer installation.

> **Modified KISS Principle: Keep It Simple and Safe**
> —Rodger M. Ewbank

Plant designs should be based on a risk assessment that considers the process and the site in detail as well as all of the principles of inherently safer operation. Earlier decisions may limit the options in the design stage, but inherently safer principles can still be applied. The design step is the last step at which changes can be made at moderate cost. Once the facility is constructed the cost of modification usually increases notably.

In some situations, a risk analysis (perhaps quantitative) may be of value in selecting from the options. Use of the risk analysis can reveal both critical equipment and critical procedures. Once the critical issues are identified, design to accommodate them.

Simplicity in design is preferred. The modified KISS principle "Keep It Simple and Safe" is applicable.

Process Design Basis

To reduce the potential for large releases of hazardous materials:

- Minimize or eliminate in-process inventory of hazardous material, including inventory in the processing equipment as well as in tanks. Elimination of intermediate storage tanks will likely require improvements in the reliability of the upstream and downstream operations.
- Review dikes, impoundments and spacing for tanks storing flammable materials. A sump inside a dike facilitates the collection of small spills. Sump drains or pumps can direct material to a safe and environmentally acceptable place. See the latest issue of National Fire Protection Association (NFPA) 30.
- Review the layout to minimize the length of piping containing hazardous material.

In batch operation minimize pre-charging the most energetic chemical. Consider adding energetic material in a "semi-batch" mode. That is, add most of the ingredients initially, then add the energetic material under flow control with a safety interlock to stop the feed as soon as the Critical Safe Operating Parameter (frequently temperature or pressure) approaches the limits of the safe operating window. Consider a physical limit (pipe size, orifice, limited pump capacity) to limit available energy. Low temperature can be dangerous if the energetic material "pools" unreacted in the kettle and then the reaction initiates. The "pooled" material could have enough potential energy to result in catastrophic releases.

When dealing with flammable materials, the selection from the inherently safer design options may vary according to the site and process. For example,

- Use nonflammable materials.
- Inert the vessel.
- Design the vessel to withstand the pressure generated.
- Install explosion suppression.
- Install relief panels (directed to a safer location).

Equipment

Although many engineers provide only the minimum adequate vessel design to minimize costs, it is inherently safer to minimize the use of safety interlocks and administrative controls by designing robust equipment. Passive hardware devices can be substituted for active control systems. For example, if the design pressure of the vessel system is higher than the maximum expected pressure, an interlock to trip the system on high pressure or temperatures may be unnecessary.

Creating a strong system that constitutes a passive design requires a complete knowledge and characterization of the potential overpressure scenarios. This requires knowledge of the chemistry outside the design conditions to evaluate effects of loss of utilities and the loss of control systems.

Systems with a passive design fully withstand any overpressure and the yield point stress of the system is not exceeded. When an overpressure stresses the vessel system, the metal returns to its normal crystalline state after stretching. Systems designed to "bend but not break" slightly exceed the plastic region of the metal and are deformed (hardened). The vessel is then actually made stronger by this process; however, a new hazard is that the vessel will not stretch and will usually burst if the scenario is repeated. Thus, vessels subjected to "bend but not break" conditions require more frequent inspections for deformation and integrity. A truly passive design is not only safer, it is more cost effective when the lifetime test and inspection requirements are considered.

A passive design must include all hardware elements of the system. Little is gained if containment is lost when pipes, joints, or instruments fail due to overpressure.

In designing the process and equipment, use chemical engineering principles to minimize the accumulation of energy or materials and to contain the energy and materials:

- Specify design pressures high enough to contain pressures generated during exothermic reactions and avoid opening the relief valve and/or rupturing the vessel.
- Use physical limits of pipe size, restrictive orifices, and pump sizing to limit excessive flow rates.
- Consider the incident avoidance benefits of reliable equipment when specifying hardware.
- Use inherently safer strategies when establishing company design standards, guidelines, or practices.
- Use gravity flow in plant layout where feasible to minimize the need for pumps or solids handling equipment for hazardous materials. Conduct a hazard review to assess the effect of layout on potential spills.
- Review injection points for erosion concerns. Design for lower velocities.
- Use materials of construction that enhance inherently safer operations. Corrosion leads to leaks; incompatible materials can lead to unwanted reactions.
- Use materials with low corrosion rates for the process.
- Use the right alloy for the job (more expensive is not necessarily better).
- Use materials that are applicable over the full range of operating conditions such as normal, startup, routine shutdown, emergency shutdown, and draining the system. For example, carbon steel may be acceptable for normal operating conditions but may be subject to brittle fracture at low temperatures under abnormal conditions (as in the case of a liquefied gas). Cold water, of less than 60°F, during hydrotest may cause brittle fracture of carbon steel.
- Avoid materials that crack or pit; uniform corrosion is safer than nonuniform corrosion patterns.
- Avoid incompatible materials that could come into contact in abnormal conditions.
- Do not use copper fittings in acetylene service, or titanium in dry chlorine service. These principles also apply to gaskets, lubricants, and instruments.

When reviewing the materials of construction consider external corrosion concerns. Chloride stress cracking of stainless steel can be initiated by insulation capturing chlorides or insulation that contains chlorides (stainless steel should be primed). A weather barrier is needed. The principle here is to understand the potential hazards and their mechanisms.

If possible, eliminate inherently weak equipment like sight glasses, hoses, rotameters, bellows, expansion joints, and most plastic equipment. The spare parts consumption from the shop and warehouse will indicate what is wearing out.

Minimize contamination via fewer cross-connections and fewer hose stations. Minimize the number of hoses required in loading/unloading facilities. Cross-contamination, sometimes even from catalytic amounts of material, can result in undesired hazardous reactions. To prevent contamination due to rainwater and spills, consider storing a material that can react vigorously with water under a roof.

Flexible connections should never be used as a cure for improper piping alignment and piping support concerns. Figure 4.4 illustrates both good and poor piping alignment. Where expansion joints are required in piping systems containing toxic materials, consider using double-walled expansion joints with pressure indication between the two walls for leak detection. All welded pipe is preferable to flanged piping, and threaded piping should be avoided for flammable and toxic materials.

Design pipe to fit properly.

Consider a weak roof seam for API tanks; if the tank is going to split under internal pressure, the roof seam should fail, not the bottom seam. The weak roof seam must be specifically included in the specifications and the mechanical design must address the issue. This emphasis is made because smaller tanks (less than 50 feet in diameter) manufactured under API 650 will not automatically have a weak roof seam.

Pinch valves that have no packing to leak can leak if the tube breaks.

There are trade-offs for magnetic drive and canned pumps versus centrifugal pumps with double-mechanical seals. The former have no seals to leak, but need active interlocks to prevent high temperature for temperature sensitive materials. Similarly, diaphragm pumps, that have

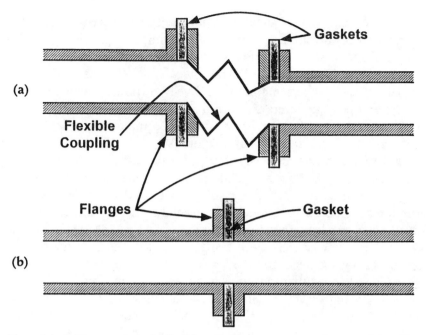

Figure 4.4. (a) Improper use of a flexible connection to compensate for poor piping alignment. (b) Proper piping design and alignment.

no shaft seals to leak, have potential for process leaks out of the exhaust line and air leaks into the process on diaphragm failure. Double diaphragm pumps need two failures before material leaks out the exhaust line.

Process Control

Many of the inherently safer design aspects discussed here appear in *Guidelines for Safe Automation of Chemical Processes* (CCPS, 1993b). It makes excellent reading for greater depth and treatment of inherently safer/process control concepts.

The ultimate goal of inherently safer design is elimination of all hazards with no need for controls. However, some control systems are always required. The working logic of a specific control system can make it inherently safer than other alternatives.

Design and development of inherently safer process chemistry and physical treatment may be the most economical way to eliminate a hazard. A hazard that is avoided by inherently safer design does not require active control. When designing a chemical or physical process, keeping the process within safe operating limits requires identification of all the system phenomena. Inherently safer designs ensure that the normal or quality operating limits are well within the safe operating limits; safe operating limits are within the instrumentation range; and the instrumentation range is within the equipment containment limits, as shown in Figure 4.5 and Figure 6.3.

If active controls are needed to prevent the process operating parameter from reaching a hazardous condition, an inherently safer design will specify the desired operating conditions to provide adequate time for controls to function before reaching the equipment limits. Calculating the adequate time requires knowing the speed of the upset, the response time of the device being controlled, the lag time of the control sensing element, and the lag time of the control final element (e.g., valve). For example, an upset in a feed to a tank could lead to an overflow on high level. An inherently safer design provides enough time for the tank level control to sense the upset and to take corrective action on the flow into or out of the tank before the tank overflows.

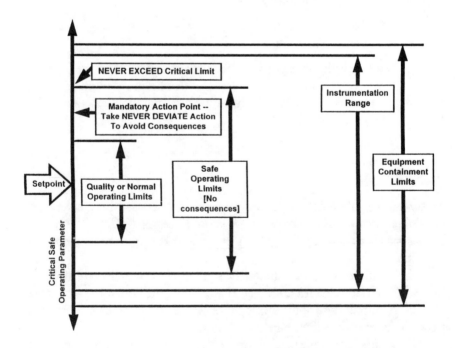

Figure 4.5. Operating ranges and limits.

For such a tank, the maximum setpoint for the level must be reduced to allow adequate response time.

For some processes, human intervention is included in the response to an upset. See Section 6.5 for a discussion of the human response time in process control.

Response time considerations lead to the establishment of never exceed critical limits and mandatory action points. The never exceed critical limit is that point at which unsafe consequences will occur. The mandatory action point is a value set low or high enough to allow time for the instrumentation or human controls to prevent the hazard. One example of response lag is a system's ability to maintain a heat balance. In reacting systems, control is lost when the system generates heat faster than it can be transferred to the surroundings. This excess heat produces an immediate pressure rise in systems containing volatile materials. In other systems, the reactants decompose to gases, resulting in a pressure rise. The mandatory action temperature limit is set low enough to allow time for the response to prevent loss of containment from high pressure.

Mandatory Action Points and Never Exceed Limits Have Two Purposes:
   —Prevent loss of containment due to exceeding equipment design strength.
   —Provide time for instrumentation or human controls to work.

The mandatory action points must not be disregarded to produce more product in the same equipment. For example, reaction rate can be accelerated by adding catalysts, increasing temperature, increasing pressure, or changing phase. An inherently safer plant provides a documentation and management of change system for the mandatory action points, never exceed limits and fail-safe positions of the emergency shutdown and process control systems. The documentation could be on the piping and instrument diagrams (P&IDs). When a process change is contemplated, good documentation can be used to assure that the never exceed critical limits are understood and are not violated by the change. The current operating team must be able to observe the heat and pressure balance controls from the Basic Process Control System (BPCS) and the Safety Interlock System (SIS).

Basic Process Control System (BPCS)
and Safety Interlock System (SIS)
There are few chemical plants that are so forgiving that a control system or a safety interlock system is not required. Process engineers provide controls to assure product yield and quality and maintain safe operating conditions. This type of control system is a BPCS. The BPCS acts to alarm and moderate a high or low operating condition specified by the normal operating limits within the never exceed critical limits. The SIS is provided to shut down or otherwise place the process in a safe state if the BPCS fails to maintain safe operating conditions. A BPCS should not be used as the sole source of a process safety shutdown.

Inherently safer SIS should be fully independent of the process control system logic residing in the BPCS. Common-mode failure can result if the BPCS and SIS share components. Operators should always get notification from the BPCS and the SIS as the mandatory action points are approached. Switches should be avoided because they do not give an indication that variables are approaching their trip set points. When switches seize, operations is unaware of the failure until a test or actual demand occurs. An example is a high temperature light on an automobile. It gives no indication of an abnormally high temperature until the temperature reaches the high alarm point. A temperature gauge gives advance warning before the alarm point. The gauge pointer provides feedback of its functionality by normal fluctuations. In another example, low level switches were installed to trip fuel to a steam boiler. In a shutdown for an unrelated cause, an engineer noticed that the level switches did not show low level when the water was drained out of the boiler. The investigation found the insulation on the wires to the switches had melted from the boiler heat creating a short circuit that prevented the low level switch from tripping—the wrong wire had been installed. An inherently safer design could use temperature sensors with indication available to the operator. Further SIS design help is available from ISA (1996).

> **Avoid common-mode failures between the BPCS and the SIS.**
> **Do not share components within a BPCS and SIS.**

When choosing SIS input variables, attempt to use a direct reading, not an indirect reading; if pressure is important, measure pressure directly rather than inferring it indirectly from temperature. This choice eliminates the lag time in processing which occurs when an indirect variable is chosen. Direct readings also eliminate potential errors in the inferred relationship between the variables. It is appropriate to show backup information that supports the directly sensed variable providing additional confirmation of proper SIS function. This can be done using analog or digital I/O devices.

> **Whenever possible, sense the process variable to be controlled. For example, for pressure control—sense pressure.**

Programmable Electronic Systems (PES) are designed to control the desired reactions or physical process using a BPCS or produce a shutdown using a SIS when the never exceed limits are reached. The programming of a PES must be well documented and easily discernible to the operator. CCPS (1989a) Chapter 6 deals extensively with redundant input, logic solver (PES) and output device functions. Proper configuration of these elements which make up a BPCS or SIS can produce an inherently safer control system. Redundancy within the input elements is inherently safer if the inputs are different variables where the phenomena is predicted equally by any of the variables sensed. Redundant logic solver configurations provide duplicate microprocessor functions. Output variables become inherently safer when like elements are duplicated, for instance, when solenoid valves and dissimilar elements like block valves and control valves are used.

Signal gathering is important in BPCS and SIS design. Routing all data through a single data link input/output (I/O) card can impair BPCS and SIS integrity by producing a common mode failure. An inherently safer design arranges inputs and outputs to be independent.

> **Multiple I/O card systems provide an inherently safer design.**

Trips should deenergize circuits and not require the shutoff/shutdown circuit to energize. Process creep can be documented by

BPCS/SIS functions that produce a trend analysis of input and output functions. Graphic video displays set-up to indicate process conditions and valve positions assure that the status of the process conditions is announced to the board operator, who may then verify the conditions with the field operator.

Startup, Shutdown, Tests, Inspections
Process control and safety shutdowns must be provided during all modes of operation, not only in the "RUN" mode. Other modes will require a BPCS configured for the mode operating algorithm and very likely a different set of safety interlocks must provide appropriate protection. Hardwire devices, like timers or software logic, can be used to actuate the SIS pertinent to the operating mode.

Use an appropriate SIS for each operating mode.

Increased operator awareness can be prompted by use of rate of process variable change, at absolute process variable milestones. For example, devices can indicate or be set to alarm if the rate of temperature rise is 5°C/minute for the 10 minutes preceding a process temperature of 100°C.

Provide the operator all the feedback information and controls needed to maintain the process within the never exceed critical limits.

An inherently safer control system design will, where possible, include options for on-line reliability tests and inspections. Within digital devices, this is done using watch-dog timers to assure the signals are processed.

It is inherently safer to do tests on a system without disconnecting the leads.

Design of the SIS must include provisions for testing. Consider the special testing situations that will exist. Do not test the SIS by creating a challenge in an operating plant. Do not test the high level shutdown by raising the level. Turbine tests are one exception to doing tests when the system is not in a fail-danger state. Turbine mechanical overspeed trips must be tested by increasing the speed since centrifugal force is needed to activate the trip.

Special (proof) test programs of safety devices always produce an uncertainty. During the test procedure, an incorrect adjustment or neglecting to reconnect the leads can compromise the design premise of the safety control. It is inherently safer to test an SIS as a requirement (permissive) for normal startup in a non–fail-to-danger scenario rather than to test it in a special test program. Since the interlock must function to start up, there is greater confidence that it will be tested. Permissives in combustion control systems are an example where test of the SIS is an integral part of the start procedure. When appropriate, SISs may require testing with the plant still running. In these instances a fully functional backup system or equivalent SIS is required.

- **Never test a SIS on-line without a fully functional backup.**
- **Never operate using the SIS as a BPCS.**

Alarm Management Ergonomics

With the ability to make every signal into an alarm in a BPCS or SIS, operator information overload is a genuine safety concern. Inherently safer alarm design will produce an alarm hierarchy. This design permits training operators to understand the importance of the never exceed critical limit alarms. BPCS/SIS digital and any analog alarm displays should be grouped to be readily identifiable by color, physical position, and distinctive sound annunciation. It is inherently safer to display an on-screen list of potential action alternatives. Ergonomics is further discussed in Chapter 6.

Humans require time to react to process alarms and control requirements. Reaction time must always be considered early in the plant design. It is inherently safer to decide early in process design what administrative controls the operator will be assigned to activate for safety control. Requiring periodic operator interface to the process system relieves boredom and heightens interest in knowing the current condition of the process. See Sections 6.4 and 6.5.

> **Analog and digital BPCSs and SISs require much less reaction time than humans.**

Supporting Facilities

Considerations regarding supporting facilities such as utility systems, water treatment, and heat transfer include:

- Do not allow nitrogen or air supplies to overpressure tanks or vessels. Tanks and vessels could be designed to withstand the air and nitrogen header pressure. Another solution is to install a pressure relief valve downstream of a pressure reducing station sized to relieve the entire flow on failure of the station.
- Use water or steam as a heat transfer medium rather than flammable or combustible oils (Kharabanda and Stallworthy, 1988; Kletz, 1991b).
- Use high-flash-point oils, or molten salt if water or steam is not feasible (Dale, 1987; Kletz, 1991b).
- Seek alternatives to chlorine for water treatment and disinfecting applications. For example, sodium hypochlorite has been used both in industrial and municipal water treatment applications (Somerville, 1990). Other alternatives include calcium hypochlorite, ozone, ultraviolet radiation and heat treatment (Negron, 1994; Mizerek, 1996).
- Use magnesium hydroxide slurry to control pH, rather than concentrated sodium hydroxide (Englund, 1991a).

Other Design Considerations

A design that is "not buildable" invites change during construction. The change may not be well designed due to perceived time pressures. Also, a change may not be recognized as significant. In the Kansas City Hyatt Regency Skywalks incident, the initial design of a hanging support was "not buildable" and the contractor improvised an alternate design. The second design was not strong enough to withstand the forces involved when a number of guests started dancing on it. The Skywalk fell resulting in injuries and death to many of the guests (Petroski, 1985). Actually, there are two lessons to be learned:

- Be sure the design is buildable
- Review design and construction changes for safety (follow "Management of Change" procedures).

Siting, both location and layout, is critical for inherently safer plants. Evaluate siting with respect to the risk imposed by the process on the population, environment, adjacent facilities, and community. These evaluations apply to small revisions as well as major new processes.

Layout considerations include avoiding crane lifts over operating equipment, especially equipment that contains hazardous materials, and, in particular, hazardous materials held under pressure or high temperature. Allow space for maintenance access without damage to other equipment. Apply risk management concepts to the layout and siting.

Additional references on inherently safer features include:

- *Guidelines for Safe Storage and Handling of High Toxic Hazard Materials* (CCPS, 1988a)
- *Guidelines for Vapor Release Mitigation* (CCPS, 1988b)
- Chapter 2 of CCPS's *Guidelines for Engineering Design for Process Safety* (CCPS, 1993a) provides an Inherent Safety Checklist. This checklist has been adapted and is provided in Appendix A.

Additional considerations presented in subsequent chapters include:

- The facility design must consider decommissioning issues, as discussed in Section 4.6.
- The equipment design must be compatible with the human factor requirements, as presented in Chapter 6.
- Design alternatives should include inherently safer reviews. Possible methodologies are presented in Chapter 7.

4.5. Operations, Maintenance, and Modifications

A primary objective of any safety program is to maintain or reduce the level of risk in the process. The design basis, especially inherently safer features that are built into the installation, must be documented. Management of change programs must preserve and keep the base record current and protect against elimination of inherently safer features. For identical substitution, the level of risk in the process is maintained, but the opportunity to improve the margin of safety is lost. Be careful to maintain inherently safer features.

For example, increasing a valve size or installing a larger pump could result in high pressure in a vessel, thus increasing the risk of a release. Sanders (1993) presents a number of examples of changes affecting the safety of a plant.

When making changes, look for opportunities to make the system inherently safer. For example, when replacing worn, corroded equipment, debottlenecking, or implementing a process improvement, look for opportunities to reduce the level of risk in the process. Look for chemistry changes as well as equipment changes. However, be aware of the need to retrain operators and mechanics when changing to an inherently safer facility.

> **We ought never to change our situation in life without duly considering the consequences of such change.**
>
> —Aesop

The safety status of the process should be periodically reviewed against the guiding principles for the original design. Monitoring of add-ons can detect potentially dangerous modifications. Process hazards analysis or process safety audits are useful tools for this review. Documentation of inherently safer principles is critical to ensure that future changes don't nullify the positive features of the initial installation.

Inherently safer principles also apply to mechanical integrity requirements. Design plant modifications for ease and reduced risk in inspections, code compliance, and maintenance. For example, if one can check the pipe thickness from a platform, it is more likely to be done than if a crane is required.

During process hazard reviews, evaluate each safety critical device or procedure to see if the device or procedure can be eliminated by applying inherently safer principles. Consider the existing plant, the next plant, and the plant after next.

Inherently safer design features include the use of:

- Rising stem valves to indicate valve position.
- Mechanical connections (or disconnections) for blanking, draining, cleaning and purging connections so that maintenance activities cannot be started without first disconnecting lines

that might add hazardous materials to the equipment. One example is to route a nitrogen line across or through a manway so the vessel cannot be entered without disconnecting the nitrogen line.
- Accessible valves and piping to minimize errors.
- Adequate spacing to avoid crowded vessel access.
- Logical numbering of a group of equipment. Figure 4.6 is an example of poor assignment of equipment numbers for pumps.

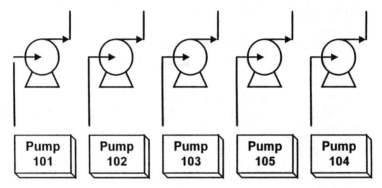

Figure 4.6. An example of poor assignment of equipment identification numbers. Where is Pump 104?

- Logical control panel arrangements. Figure 4.7 is an illogical arrangement of burner controls on a kitchen stove.

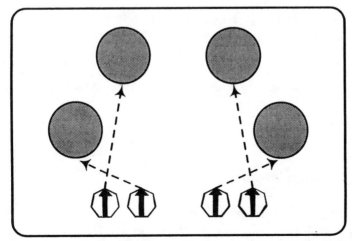

Figure 4.7. An illogical arrangement of burner controls for a kitchen stove (from Norman, 1992). Note: The author who drew this figure had a difficult time getting it "right." His instinct was to connect the controls and burners in a logical fashion.

- Organized material safety data sheet information for the specific process highlighting potential trouble spots for maintenance materials, including materials of construction, lubricants, or packing.

Consider inherently safer procedures such as:

- Using checklists.
- Setting priorities.
- Scheduling to minimize cross-contamination and clean-ups.
- Recording a value for a process variable rather than just a check mark.
- Using a second person to double-check.
- Using clear procedures to ensure wearing of appropriate safety equipment (avoid "gray" areas of interpretation). Avoid having complex criteria for determining the recommended personnel protective equipment (simple is easier to understand and to follow).

Materials of construction must be correctly specified and installed during maintenance. Sometimes people use what they have, not what was specified.

For operability and maintainability, consider if the total number of devices and the number of types of devices can be reduced. This reduces the risk of installing a wrong device. The tradeoff is that the standard device may be larger than needed for inherently safer operation at one service location.

Spacing standards and process review criteria must apply to modifications and changes in the same way that they apply to the original design.

Many maintenance risks are related to unclear and unsafe procedures. See Chapter 6. Additional information regarding management of change can be found in Chapter 7 of the CCPS *Guidelines for Technical Management of Chemical Process Safety* (CCPS, 1989b).

4.6. Decommissioning

The design and implementation of inherently safer chemical processes includes consideration not only of the people operating and maintaining the plant while it is in operation, but also the safety of those who may later use the plant for other purposes. This includes dismantling of process equipment, reuse of the site, or impact by chemicals left behind in the plant or left in the soil or groundwater at or near the plant site. The process equipment and ancillary equipment must be removed or at least left in a safe condition. Wastes, landfills, soil, and groundwater must be left in a safe condition and possibly be monitored and remediated. In the United States and in most industrial countries, closure and postclosure activities are legally regulated. Equipment, which includes instruments and lines, must be thoroughly cleaned before abandonment or reuse.

Example 4.6.1

A 50-gallon stirred pot reactor was used for the production of sodium aluminum hydride, which reacts exothermally with water with enough heat to cause the hydrogen that is released to explode. The reactor was emptied, cleaned thoroughly (by report), and then placed in an outdoor "surplus equipment yard" with the nozzles open to "weather." About one year later, a maintenance man was ordered to clean up the reactor in preparation for reuse. He was told to put on full protective fire gear before opening the vessel. He did not put on the fire gear and proceeded to open the vessel and hose it out with a fire hose. An explosion resulted when water dislodged crusted-over sodium aluminum hydride trapped in a nozzle. The worker was burned, requiring a two-week hospital stay and several months of recuperation.

Attention must be given to the long term protection of people or the environment from hazards from abandoned equipment. Equipment that meets the criteria for disposal in a landfill—that is, it has been properly cleaned—may not be suitable for other uses. Problems such as the one related in the following example can be avoided by making the abandoned equipment unusable if it cannot be cleaned adequately.

Example 4.6.2

Reactors equipped with heavy agitators used for tetraethyl lead manufacture during World War II were disinterred from bomb rubble and were found by the people who dug them up to be ideal for processing fish paste for human consumption. The reactors were washed, but this did not prevent poisoning a number of people.

Example 4.6.3

Heavy-walled drums once used for lead antiknock chemicals have been used for water storage or as barbecue pits, with subsequent risk to the user from residual toxic material. Equipment from the industry cited has for many years been cleaned, cut up, and sent under supervision to steel mills for recycle to eliminate the possible misuse of scrapped containers.

There is a temptation for managers to delay the cleanup of decommissioned or abandoned plants as long as possible. However, experience teaches that there will never be a time for chemical plant closure activities that will be less expensive than immediately after the plant is closed. That is the time the most people who know how to handle the materials and who know where materials are located are available. That is the time when units are still intact, decontamination can be most easily performed, decontamination procedures are well-known, and decontamination equipment is available. That is the time when waste disposal contracts that cover the materials in the plant are still open or can be reopened easily. That is the time when design documents, waste manifests, maintenance records, and other files are most likely to be readily available. That is the time when equipment has probably not corroded to the point that it can't be handled safely, that valves will open, that nuts are not frozen, and that instruments are in working condition. That is the time when you are least likely to encounter tanks, drums, or whatever with contents nobody can identify without an expensive analytical investigation. Generally, if there are contaminants in the groundwater, the longer remediation is delayed, the larger will be the area to be cleaned up.

Plants can be made inherently safer for closure by eliminating or at least reducing the amount of hazardous material to be removed from the soil, the groundwater, and process equipment, and by properly constructing and maintaining disposal facilities such as landfills. For example, some new chemical plants are being designed to "start one floor up" by making the lowest operating level one story up and keeping all utility and sewer lines above ground with the ground level paved and curbed. Most new plants are designed with double-containment for sewers with provisions for early detection and correction of leaks. Many existing plants have been retrofitted with this more secure sewer system. Extraneous lines and equipment that are removed from service because of process modifications will not present a "mystery" or a problem at the time of closure if they are removed when the modification is made; management of change procedures should require removal of extraneous equipment and lines.

5

Transportation

Addressing transportation risk at various life cycle stages can increase the inherent safety of the overall operation. For example, this may mean minimizing the number of railcars sitting on track within or outside the plant, selecting suppliers close to the plant, and so on. This chapter provides an overview of inherently safer concepts applied to transportation.

In choosing the location of a new plant or in assessing risks related to an existing plant, transportation risk must be considered. The design of new chemical processing units should include at the earliest opportunity a qualitative or quantitative risk assessment of the whole system including production, use, and transportation in order to minimize overall risk. Risk assessments of existing processes, including an assessment of transportation risk, may result in the conclusion that the process should be moved to reduce risk. Process risk assessment techniques are available in numerous references (CCPS, 1989a) and the analysis of transportation risk is documented in *Guidelines for Chemical Transportation Risk Analysis* (CCPS, 1995b).

The assessment of transportation risk must include consideration of the capabilities, equipment, and practices of both raw materials suppliers and customers. Some apparently attractive options may need to be discarded because of what suppliers of raw materials or customers can or will do. In some cases, suppliers' or customers' capabilities may need to be upgraded to acceptable standards.

The chemical industry and carriers are working together to improve both transportation safety and response to emergencies. The American Association of Railroads (AAR), among others, is working with the Chemical Manufacturers Association (CMA) Inter-Agency Task Group to recommend improvements in equipment, routing, and procedures to enhance safety. The Chlorine Institute has for years acted to provide inherently safer transport of chlorine. The CMA Responsible Care®, Chemtrec®, and Transcaer® programs have resulted in significant improvement in transportation safety and emergency response.

There are many regulations governing the transportation of chemicals, and any evaluation of transportation risks and options must include consideration of those regulations. In addition, some companies have policies that require going beyond legal requirements for specific materials.

5.1. Location Relative to Raw Materials

It may be possible to reduce or eliminate transportation risk by locating the plant where hazardous raw materials or intermediates are produced if the risk from transporting the raw materials or intermediates outweighs the risk of transporting the final product. Locating the starting and ending plants at the same site will probably provide additional opportunities for risk reduction by inventory reduction. Of course, increasing the number of facilities at a particular site may increase the overall risk at that site.

Example 5.1

A plant produced methyl methacrylate by reacting hydrogen cyanide with acetone to produce acetone cyanohydrin followed by further processing to produce methyl methacrylate. The hydrogen cyanide was produced at another site and was transported to the methyl methacrylate plant by railcar. A hydrogen cyanide plant was subsequently installed at the methyl methacrylate plant site to eliminate the need for shipping hydrogen cyanide or acetone cyanohydrin.

Example 5.2

A company produced bromine in Arkansas and brominated compounds in New Jersey. A risk assessment resulted in a recommendation to consider the transfer of the bromination processes to the bromine production site in Arkansas. Economics and the decrease in risk justified such a transfer and it was done. Although safety was not the only consideration, it was an important factor in this decision.

5.2. Shipping Conditions

The physical condition and characteristics of the material shipped should be considered in transportation risk assessments on a case-by-case basis. There may be options available to reduce transportation risk by reducing the potential for releases or the severity of the effects of releases. A few possible ways of improving safety by modifying conditions are:

- Refrigerate and ship the material at atmospheric pressure or at reduced pressure.
- Ship concentrate to reduce the number of containers, then dilute the concentrate at the user site.
- Ship and use the material or a substitute in diluted form; e.g., aqueous ammonia instead of anhydrous ammonia, or bleach instead of chlorine.
- Ship and use intermediates rather than raw materials.

5.3. Transportation Mode and Route Selection

Select a transportation mode to minimize risks to the extent practicable. Drums, ISO tanks, tank trucks, rail tank cars, barges, and pipelines offer tradeoffs in inventory, container integrity, size of potential incidents, distance from supplier or customer, and the frequency of incidents.

Barges may have fewer accidents than tank trucks, but the severity of a major release from a barge may be great enough to make the higher potential accident rate with tank truck shipments the better choice.

The transportation mode used will affect the shipper's options with regard to the selection of the routing of the shipment. Using certified drivers to ship full truckloads of drums, ISO tanks, and tank trailers may allow the shipper to specify routes to avoid high risks. The time of day and duration of travel can also be specified. The shipper generally has little control of the routing of less-than-full truckloads mixed with other freight unless routing is specifically spelled out in the contract.

Railroads choose the routing of rail tank cars and should be contacted to see if arrangements can be made to minimize risks. Improved tracking of rail shipments by the railroads should reduce hazards such as long-term storage of chlorine tank cars on spurs adjacent to residential areas. The routing of barge shipments is essentially fixed by the location of the shipper and the receiver, and there is generally no choice of routing with a pipeline unless one wishes to install a dedicated pipeline. Data on accident rates by mode and references are given by CCPS (1995b) and can be used to select the safest alternative.

Carriers and the chemical industry are working together to improve transportation safety. The American Association of Railroads has agreed to designate routes that handle 10,000 loads per year or more of chemicals as "Key Routes." Routes designated as "Key Routes" will receive upgraded track, enhanced equipment to detect flaws in equipment or in trackage, and lower speed limits.

"Just-in-time" supply of materials may affect the mode of transportation and could increase risks from the material. For example, drums of a chemical could be stockpiled near a user and not be under the level of control that could be provided by either the supplier or user if the inventory were maintained in a storage tank at one or the other facility. This type of risk should be included when contemplating "just-in-time" shipments.

5.4. Improved Transportation Containers

Transportation risk can be reduced by applying inherently safer design principles to transportation containers. Some examples of design improvements follow.

- The CMA Inter-Agency Task Group has proposed many initiatives to improve transportation safety. For example, the shipment of environmentally sensitive material in General Purpose rail cars will be phased out by the year 2000 and DOT Specification 105 pressure cars will be used instead.

- Remote controlled shutoff valves can reduce the severity of incidents.
- Thermal insulation can be used to maintain lower temperatures in the containers and to provide improved protection from fire.
- Tank cars, trailers, and other containers can be specified without bottom outlets or be provided with skid-protection for bottom outlets.
- All rail tank cars must be equipped with roller bearing wheels.
- Using a container designed for the maximum pressure that the contents can generate from ambient conditions will eliminate the need for refrigeration of the container for safety.
- Overpackaging can be used to provide maximum protection; for example, use DOT Specification 105 tank cars instead of general purpose cars.
- Rail cars for chemicals have been designed with "shelf" couplers or double shelf couplers and reinforced tank ends to reduce releases from accidents.
- Some barges have double hulls.
- "Low-boy" trailers are used for truck movement of ISO containers. Low center of gravity trailers are also available for tanks. The lower center of gravity reduces the risk of a turnover.
- Baffles can be used in large containers to improve stability.
- Nonbrittle containers can be used to improve resistance to impact or shock damage.

5.5. Administrative Controls

In addition to improving safety during transportation by optimizing the mode, route, physical conditions, and container design, the way the shipment is handled should be examined to see if safety can be improved. For example, one company tested to determine the speed required for the tines of the forklift trucks used at its terminal to penetrate its shipping containers. They installed governors on the forklift trucks to limit the speed below the speed required for penetration. They also specified blunt tine ends for the forklifts.

A program to train drivers and other handlers in the safe handling of the products, to refresh that training regularly, and to use only certified safe drivers is another way of making transportation inherently safer.

6

Human Factors

The guiding premises for making systems inherently safer against human error are:

- Humans and systems designed and built by them are vulnerable to error.
- Existing facilities can contain many traps to cause human error.
- Designers can provide systems to allow error detection and to effect recovery before the error becomes serious.

CCPS (1994a) and Lorenzo (1990) discuss human error in detail. They

- offer guidance on the underlying theories of human error and the role of performance influencing factors,
- summarize techniques for human error analysis and quantification,
- suggest methods for collecting data on human error, provide example case studies, and
- discuss a systematic approach to human error.

The tools in CCPS (1994a) can be used in each stage of the chemical process life cycle to help evaluate the tradeoffs involving human factors between various options. In many cases, low cost options in design can make the operations inherently safer from a human factors perspective.

Additionally, the tools in CCPS (1994a) can be used to build inherently safer human systems for each stage of the chemical process life cycle. Human systems include

- appropriate training,
- reviews,
- audits, and
- error correction cycles.

Well-designed human systems can produce inherently safer plant designs and operating procedures. If we understand how humans work and how human errors occur, we can design better systems for managing, supervising, designing, reviewing, training, auditing, and monitoring.

The discussion in the following sections is not comprehensive; it is intended to provide examples. The inherently safer design strategy (see Section 2.6) used in most of the examples is *Simplify*.

6.1. Overview

From a human factors perspective, the chemistry of the process can be made inherently safer by selecting materials that can better tolerate human error in handling, mixing, and charging. If a concentrated reagent is used in a titration, precision in reading the burette is important. If a dilute reagent is used, less precision is needed.

Likewise, the equipment can be made inherently safer for human factors by

- making it easier to understand,
- making it easier to do what is intended,
- limiting what can be done to the desired actions.

The "process" includes more than the equipment and the chemistry. It includes the systems of training, supervising, and providing tools to the people who operate and maintain the plant. If one designs certain features into the system of training and supervising the people, they will operate the plant in a safer way. For example, avoiding fatigue by optimum shift rotation is inherent in the operating system. It is inherently safer if such features are incorporated into a **SYSTEM**. One cannot depend on the operators and mechanics learning such things by chance.

> **Features that help the operators and maintainers run the plant in a safer way are inherently safer if they are incorporated into a SYSTEM.**

As new equipment becomes available and insights about ergonomics and human factors become available, we should review new and existing facilities to optimize the person–machine interface.

One may argue that many techniques listed in this chapter are simply "safer," rather than "inherently safer." While there is some overlap, a technique becomes inherently safer if it is systematically designed into the process, equipment, and people systems in a unit, plant, or company. If only one or two people practice it on their own, it is simply "safer," but it is not an inherently safer system

New facilities should be reviewed for ergonomics and human factors issues during design, construction, and startup. Existing facilities should be reviewed periodically for opportunities to improve human factors in an inherently safer way.

6.2. Operability and Personnel Safety

There may be well-run facilities in which the operators are doing their best to "be careful" with facilities or systems that could be redesigned to be inherently safer. These facilities will be inherently safer if designed for operability. Note that inherently safer human factors features can reduce risk of injury to employees (improved personnel safety) and can reduce risk to the process from the worker (improved process safety).

Ergonomics should be applied in layout of equipment, valves, controls, and anything else that the operating and maintenance personnel need to access. Designs that avoid bending, climbing, and stretching are inherently safer than designs that require them.

Consider a task analysis for "Do-ability." Can the operators do what we have asked them to do? Tasks and facilities should be designed with knowledge of ergonomic considerations and performance shaping factors (PSF) so that operator reliability can be designed into the task.

Designs and systems should minimize potential harmful exposures in both normal and emergency operations. This consideration affects the location of normal and emergency drains and vents.

Example 6.1

One six-floor unit was designed with the expectation that the operators would catch drainings from sample lines and filters in a bucket and carry the bucket down the stairs to pour it into a tank. Since the streams contained a smelly, irritating organic acid, the operators found it easier (and safer, for the one highest in the structure—but not for anyone below) to flush the drained material through the floor grating with a water hose. Raincoats were mandatory for entering the facility.

To reduce employee exposure and environmental incidents, the plant deemed it necessary to add drain pans and piping under the routine samples and filter drains. The final system was inherently safer than the original. If technologically possible, eliminating the need to drain would be inherently safer than the drain pans.

Designers must include **all** the tasks in their designs. Designs for steady state operation have been reported to lack start-up or shutdown capability. Some plant designs lacked the vents and drains required to empty, flush, and clear the equipment.

> **Facilities designed with at least an outline of the operating procedures are inherently safer than processes designed without knowledge of the operating procedures.**

Designs should be based on knowledge of what the human body (and human nature) will do. Include educated operators in design reviews. The HAZOP methodology for process hazard analysis offers an excellent opportunity to identify design and procedural opportunities for inherently safer systems. After all, the "OP" in HAZOP stands for operability (CCPS, 1992). For example, a safe start-up procedure that requires the operator to walk up and down the stairs three times to manipulate valves in the correct sequence can be made inherently safer by locating the valves so that operator has to walk up the stairs only once during the start-up.

> Procedures that are easy to follow are inherently safer than those that are not.

> For inherently safer interactions of designs and procedures, include an operator trained in human factors on the design team.

6.3. Maintainability

A space station design that requires less "space walk" time is inherently safer than a design that requires more. If the astronauts don't need to go outside, they have less risk. For chemical plants, designs or operating regimes that reduce or eliminate vessel entries are inherently safer.

During the design phase, identify the human interaction with the chemical process and provide means to make that interaction inherently safer.

Example 6.2

Rail cars, tank trucks, and some reactors and storage tanks were cleaned manually by personnel who entered the vessel; fatalities occurred from unexpected or undetected low oxygen content or toxicity. An inherently safer system is a rotating pressurized water spray head that does the cleaning without vessel entry.

Elimination of filters that must be changed reduces the potential for exposure—this may require a redesigned filter or a process change that eliminates the need for the filter.

> If you don't make residues, you don't have to filter them out.
> If you don't install a filter, you don't have to maintain it.

Human factors should be considered in the location of items to be maintained and the required frequency of maintenance:

- inspection items
- calibration items (on-line, off-line, or shutdown)
- periodic replacement
- repair without shutdown

Equipment that can be reached for inspection, repair, or monitoring from permanent platforms is more likely to be inspected, calibrated, and replaced than equipment that requires climbing with a safety harness or scaffold.

Calibrating equipment usually requires disconnecting it from the process. Equipment that requires less calibration is inherently safer. A furnace oxygen analyzer is not protecting the furnace while it is being calibrated. Equipment that can function in abnormal operating conditions is inherently safer than equipment that fails in those conditions. For example, an oxygen analyzer was designed to shut itself down when the oxygen content went below 4%. While the oxygen analyzer shutdown tripped the furnace, it left the operators blind during the shutdown and delayed the restart. An analyzer that continued to show the actual concentration during the upset would be inherently safer.

Equipment could be designed so that there is only one right way to reassemble it. If it is important for a pipe sleeve to be right side up, then it could be notched or pinned so it will go in only right side up. For example, one plant found key relief valves installed backward after testing because the inlet and outlet flanges were identical. They revised the valve and piping flanges so the relief valves could only be installed in the correct orientation.

6.4. Error Prevention

To prevent errors, it is important to make it easy to do the right thing and hard to do the wrong thing (Norman, 1988). The design and layout can be clear on what should be done or it can be very confusing. Likewise, the design of the training can increase or decrease the potential for error.

Systems in which it is easy to make an error should be avoided. To reduce the risk of contaminated product and reworked batches, it is generally better to avoid bringing several chemicals together in a manifold. However, manifolding can be done safely and it may be the best design when all factors are considered.

Knowledge and Understanding

The operators and engineers need a correct mental model of how the process is operating linked to what they can see. If the operators do not understand what is happening in the process via the information available to them through the instruments and their eyes and ears, they may operate the process incorrectly—even while doing their best. For example, many people adjust the house air conditioning thermostat to a very low temperature setting in the mistaken belief that it will cool quicker—not realizing that the thermostat simply switches the air conditioning unit on or off, and a lower setting will not make it cool faster, but will only make it run longer.

Additionally, Norman (1988) discusses **knowledge in the world** versus **knowledge in the head**. With knowledge in the world, one is guided in a task by what one sees. Knowledge in the head is the memory required to do the task satisfactorily. Knowledge in the world is the sticker on the phone, "For Emergency, Dial 911." Knowledge in the head is one's memorized home phone number.

If you don't have to memorize it, you can't forget it!

For a chemical operator, knowledge in the world is the upper and lower limits for a pressure reading shown on the log sheet, along with the response to take. A log sheet with only a blank to fill in the reading requires knowledge in the head. This contrast is shown in Table 6.1.

Training requirements should be based on the task difficulty, frequency, and criticality.

Procedures must be up to date. CCPS (1996) offers guidance for writing operating procedures.

TABLE 6.1
Knowledge in the Head versus Knowledge in the World

KNOWLEDGE IN THE HEAD	
Time	**Inlet Temp, °C**
0800	
1000	
KNOWLEDGE IN THE WORLD	
Time	**Inlet Temp, °C**
	Never Exceed 100° C, start emergency cooling at 95° C, trip reactor at 97 °C
0800	
1000	

Example 6.3

One person in a unit retrieved a two-year-old shutdown procedure from his locker and tried to use it. The old procedure was significantly different from the updated procedure. If the old procedure were followed exactly, there would be more risk. If the two procedures were intermingled, there would be high temperature and catastrophic corrosion. The unit management team tore up the bootleg procedure and added the updated shutdown procedure to the controlled document system used for ISO 9002 (ISO 9000, 1994). In this document system, only the current procedure is available. Alternate systems may use an expiration date or require periodic reconfirmation.

Control Design

Avoid violation of cultural stereotypes. A cultural stereotype is the way most people in a culture expect things to work.

Example 6.4

Common examples of cultural stereotypes include:
- Light switches
  —in the USA, ↑ up to turn on
  —in the UK, ↓ down to turn on.
- Turn valves counterclockwise to open, clockwise to close, usually. (Note that "turn counterclockwise to open" is inherently safer than "turn to the left to open." For a multiturn valve with a wheel on a stem, turn to the left assumes the operator is looking at the side of wheel farthest away; the closer side of the wheel is moving to the right.)
- Hot water on the left, cold water on the right.
- Pump run lights:
  —US chemical plant: Green = Run, Red = Off.
  —US power plant: Red = Run (i.e., has power, danger), Green = Off.
  —Japan chemical plant: Red = Run, Green = Off

Example 6.5

A plant startup required heating the circulating gas to remove oxygen from the equipment. After several spurious trips, the operator noticed that all cooling fluid pumps were off (they should have been on throughout the startup), and turned them all on at the same time. Steam in the equipment condensed and the resulting vacuum imploded six vessels. The run lights and start/stop buttons were color coded differently in the control room from those outside—contributing to the cooling fluid pumps being inadvertently turned off.

What people expect varies with the culture. While "culture" may be difficult to define, we can make plants inherently safer by designing them with knowledge of the cultural stereotypes of the employees.

Example 6.6

While writing this book, the authors found the water faucets (round handles) in the committee meeting room rest room (in a famous San Francisco hotel) opened backward from expected—committee members experienced splashed clothing (Figure 6.1). We later learned that water valves with lever handles (frequently seen in hospitals or in sinks intended for disabled persons) are designed for the lever to be pushed away from the user to close the valve. Thus, the cold water valve designed for a lever will turn to the left to close. If a round handle is fitted instead of the lever . . . splash!

Therefore, we need human systems that not only install the correct valves, but can also install the correct handles!

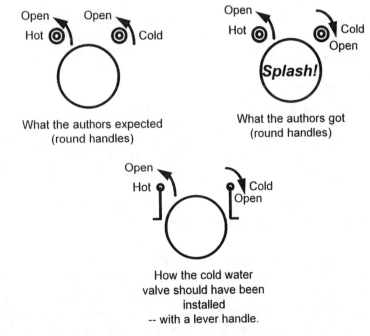

Figure 6.1. Confusing water faucets violate cultural stereotypes.

Example 6.7

While writing this book, the authors felt too hot in the hotel meeting room. Three authors opened the digital thermostat, studied the complex set of buttons, and were unable to make the room any cooler. When we asked a hotel employee, he pointed to the almost invisible switches on the bottom edge of the thermostat and said, "First, you should turn this switch to ON, and second, you should turn this switch to COOL." This incident reminded us of two things about inherent safety (Norman, 1988).

- Design the key controls so they are easily seen and are clearly labeled.
- Keep related information and controls close together, and leave out information and controls that are not needed.

If it is too complicated, no one will be able to figure it out.

Consistency

Consistency in controls and response is important; for example, upward movement on a control panel always causes the valve to open, no matter whether the valve is "air to open" or "air to close."

Example 6.8

A process waste heat boiler was damaged when the control operator inadvertently closed the cooling water valve when he intended to open it. The manual control knob turned in the opposite direction from most of the other valves in the plant—the cooling water valve was air to close—most control valves were air to open.

In an inherently safer design all of the control knobs would turn the same way to open a valve.

Consistency is important for inherently safer computer control applications. A number of empty or one-page incomplete notes were sent because [F10] was "page down" in one computer E-mail program, while [F10] was "execute/send" in another application. Critical information has been accidentally lost because [CTL-X] was "exit" in one program and "delete" in another.

Human Capability

Control systems should be designed with knowledge of the capability of human beings for required tasks. There must be a balance between totally automatic control of the process with operator monitoring versus operator control of key variables. Operators need to actually run the process enough to be able to handle it during emergencies (CCPS, 1994a).

COLOR BLIND TOLERANCE. Control system and other displays should be designed to transmit information to personnel who are color blind. Up to 5% of some male populations are red–green color blind (Freeman, 1996). If the color red is used for stop (or, closed) and green is used for go (or open), an alternate scheme should be used to transmit the same information. One approach for video displays is to design the red symbols with a red outline and black (unfilled) interior and to design the green symbols with a green outline and green (filled) interior (see Figure 6.2). Intensity (brightness) can also be used to convey some status information. Switches and indicator lights should be oriented consistently; for example, the start button should have the same position with respect to the off button. There are alternate color schemes that can be seen by most people with red–green color blindness—some examples are white, black, gray, blue, and yellow. Colored lines on a flowsheet can also be coded with dashes, dots, and crosses (this method also retains the information in a photocopy).

ALARM SHOWERS. Avoid alarm "showers" by using alarm management. Use logic to show only the alarms the operator really needs to see in

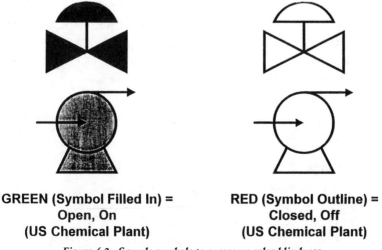

GREEN (Symbol Filled In) =
Open, On
(US Chemical Plant)

RED (Symbol Outline) =
Closed, Off
(US Chemical Plant)

Figure 6.2. Sample symbols to overcome color blindness.

the situation (see Section 4.4). Ensure that valid alarms that the operators should see are not deactivated.

Example 6.9

An alarm "shower" can occur when high pressure in the reactor trips the feed by closing a valve and turning off the feed pump. In addition to the "high pressure trip" alarm, the operator is frequently "showered" with "low feed flow," "low-low feed flow," and "feed pump off" alarms.

With alarm management, the "high pressure trip" would alarm, but the other associated alarms would be suppressed, since they are the expected result of the high pressure trip. On the other hand, a low flow condition from a different cause would be alarmed. Plants with managed alarms are inherently safer than those without since it is easy to silence and overlook a critical alarm in the midst of an alarm shower. However, the benefits should be balanced with the increased complexity and maintenance requirements.

Without alarm management, when the operator turns off the agitator, an alarm will sound for "low agitator speed." With alarm management, the "low agitator speed" would not alarm when the agitator is switched off by the operator during the proper process condition (but it would alarm for a trip of the agitator motor, a mechanical problem giving low speed, or the operator inadvertently turning it off).

From a broader perspective, the Abnormal Situation Management Consortium is working to apply human factors theory and expert system technology to improve personnel and equipment performance during abnormal conditions. In addition to reduced risk, economic improvements in equipment reliability and capacity are expected (Rothenberg and Nimmo, 1996).

FEEDBACK. Feedback can reduce error rates from 2/100 to 2/1000 (Swain and Guttmann, 1983). If the person can see that he is doing the right thing, he can be sure he did it.

Example 6.10

For a transfer from tank A to tank B, if the operators can see the level decrease in tank A and increase in tank B by the same amount, they can be confident the transfer is going to the right place. If the level in tank A goes down more than it goes up in B, the operator should look for a leak or a line open to the wrong place.

> **If you can see that you are doing the right thing,
> you can be sure you did it.**

Consider the following in control design:

- Avoid boredom—if operators don't have anything to do, they go to sleep—mentally, if not physically.
- Display corroborating or verifying information on the same display with—or very near to—the other information. Display the reading from two level sensors for the same tank on the same chart or graphic.
- Put sensibility limits on process control inputs and setpoint changes.
- Limit maximum or minimum setpoint inputs to stay in safe and quality operating regions.
- Limit the maximum step changes to setpoints to prevent upsetting the process.
- Provide bumpless (smooth) transfer and setpoint tracking for switching among automatic, manual, and cascade.
- Catch decimal errors by software or procedure. For example, have the control system logic trap and prevent setpoint changes, for example, from 6% to 61%, when a change from 6.0 to 6.1% is intended.
- Provide guidance to operators on the magnitude of a specified action to achieve a specified goal. Rather than letting the operator guess at how much to open a valve, suggest opening to 5%, then using minor adjustments to get the desired startup flow. Where needed, give guidance on how to lead or lag in changing setpoints. Advise on how long to blow a line to clear it of liquid. "Take out the guess work." Good operators will figure these tips out; document the information and make it available to all the operators (Collins, 1978).

6.5. Error Recovery

Feedback that confirms "I am doing the right thing!" is important for error recovery as well as for error prevention. It is important to display the actual position of what the operator is manipulating, as well as the state of the variable he/she is worried about.

Example 6.12

In the Three Mile Island incident, the command signal to close the reactor relief valve was displayed, not the actual position of the valve (Kletz, 1988). Since the valve was actually open, the incident was worse than otherwise.

Systems should be designed with knowledge of the response times for human beings to recognize a problem, diagnose it, and then take the required action. Humans should be assigned to tasks that involve synthesis of diverse information to form a judgment (diagnosis) and then to take action (Freeman, 1996). Given adequate time, humans are very good at these tasks and computers are very poor. Computers are very good at making very rapid decisions and taking actions on events that follow a well-defined set of rules (for example, interlock shutdowns). If the required response time is less than human capability, the correct response should be automated. Unless the situation is clearly shown to the operators, the response has been drilled, and is always expected, anticipate from 10 to 15 minutes (Swain and Guttmann, 1983) up to 1 hour (Freeman, 1996) minimum time for diagnosis.

For key operating variables, post, train, and drill the responses for Critical Safety Operating Parameters. If the process variable approaches the Mandatory Action level, the operator should take the Never Deviate corrective action (see Figure 6.3). Supervision should never criticize taking the Never Deviate action to avoid the Never Exceed limit.

> **Critical Safe Operating Parameters: Never Deviate actions prevent reaching the Never Exceed limit.**

The response should be thought out before it is required, and personnel should be trained and drilled. Old instructions would say, "If high temperature occurs, call the engineer (foreman)." Instructions should tell the operator whether to increase the cooling water flow, set off the deluge system, or evacuate! The operator should not have to "wing it" when all the alarms are going off and the relief valves are lifting.

For critical, high consequence systems, simulators are useful to practice diagnosis and correction of errors and abnormal conditions in emergency conditions (CCPS, 1994a).

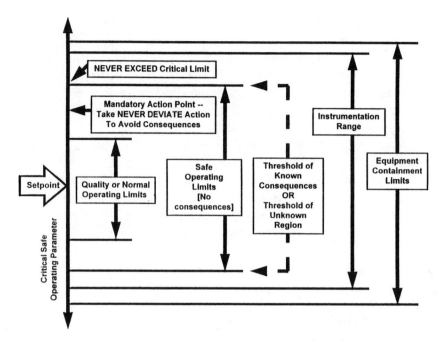

Figure 6.3. Operating ranges, illustrating Never Exceed limits.

An inherently safer operating system should also address how to use personnel effectively in response to a process upset. Without such a system, the most knowledgeable person(s) in the unit frequently rushes to attend to the perceived cause of the emergency. While this person is thus engaged, other problems are developing in the unit. Personnel may not know whether to evacuate, resources may go unused, and the ultimate outcome may be more serious. The Incident Command System should be considered for application to a process incident (CCPS, 1995c)—fire fighters and medical personnel are trained in this system. The knowledgeable person assumes command of the incident, designates responsibilities to the available personnel, and maintains an overview of all aspects of the incident. Thus, as resources become available, the process corrective actions, emergency notifications, perimeter security, etc., can be attacked on parallel paths under the direction of the incident commander.

Similarly, unit operating staffs can be trained to work together during a process upset using all the skills and resources available. Such training is part of nuclear submarine training ("Submarine!", 1992) and it is part of cockpit flight crew training for commercial airlines. This training helps overcome the "right stuff" syndrome. An inherently safer system would have personnel trained to use **all** the resources for error recovery. The test pilots in the book *The Right Stuff* (Wolfe, 1979) would rather crash and burn than declare an emergency, since an emergency was an admission that they were not in control, and therefore didn't have the "right stuff."

Error recovery by the operators is only one of several layers of protection to prevent undesired consequences (see Figure 2.1). Process and equipment designs (discussed in previous chapters) that prevent undesired process excursions are inherently safer than designs that require operator intervention. Likewise, designs that enable the operators to intervene before an upset becomes serious are inherently safer than those that do not.

6.6. System Audits

An inherently safer system should have inspection and reliability testing of safety critical systems and practices (CCPS, 1993c).

Inspect to confirm that people and equipment are really there for safety critical systems and procedures.

Develop methods to measure the effectiveness of inherent safety efforts and to provide feedback to personnel to improve performance.

Example 6.13

In one site, the project team installing a DCS (Distributed Control System) carefully developed and tested techniques to make the displays clear for red-green color blind personnel (see discussion in 6.4). The displays were effective and were applauded by the operators. However, in subsequent DCS installation projects at the same site, different project teams made no provision to make displays visible to red-green color blind personnel. This inherently less safe condition was found during a design review at one unit and at the pre-start-up safety review for another unit.

> **An inherently safer engineering system would have a mechanism to convert lessons learned from a project into guidelines and standard practices for later projects.**
>
> **It would also have an audit system to confirm that lessons learned were indeed captured.**

Audits give a snapshot in time; corrective actions are identified and executed, but the audit must be done again later to see if the system has changed. Using an analogy to process control, a continuous measurement system is desired to give warning of problems in the process management system. CCPS is completing research for a quantitative measurement system for two of the elements defined by OSHA (1992)—management of change and training. The research is defining measurable parameters, equations, and software programs to detect deviation form the design of these two elements. Once the system is set up for an organization, key sensitive indicators can be monitored to detect deviation—a new comprehensive audit is not needed. The UK Health & Safety Executive and the US Department of Energy are participating. A publication is expected in mid-1997.

6.7. Organization Culture

The performance of human beings is profoundly influenced by the culture of the organization (see discussion of the "right stuff" above). Unit/plant/company cultures vary in the degree of decision making by an individual operator. Cultures vary in their approach to the conflict between "shutdown for safety" versus "keep it running at all costs." Personnel in one plant reportedly asked "Is it our plant policy to follow the company safety policy?" In an organization with an inherently safer culture, people would say, "Our plant policy is to follow the company safety policies and standards!"

An operating philosophy that trains and rewards personnel for shutting down when required by safety considerations is inherently safer than one that rewards personnel for taking intolerable risks.

7

Inherent Safety Review Methods and Available Training

7.1. Inherent Safety Reviews

More companies are building inherently safer design principles into their process safety management systems. This can be accomplished by incorporating inherently safer design concepts into existing safety and process hazards reviews. Companies may wish to augment their existing review systems with inherent safety reviews at key points of the process life cycle. This chapter discusses a methodology to conduct inherent safety reviews at three key stages of the life cycle: during product and process development, during conceptual facilities planning early in the process design phase, and during routine operation. Many companies may elect to include the objectives and features of these reviews into their existing process safety management system. If companies decide to incorporate inherent safety into existing systems and reviews, particular attention needs to be given the subject in R&D for product and process development. This is the life cycle stage where application of inherently safer design concepts can have the greatest impact.

In a review of inherent safety awareness and practices in the United Kingdom, Mansfield (1994) reported a general lack of awareness of inherent safety principles, and their absence from company review procedures. The greatly increased quantity of literature on company inherent safety policies and procedures published during the past several years may be an indication that this situation is beginning to change. Companies which have chosen to incorporate inherent safety principles into their existing hazard management programs may want to consider highlighting or otherwise identifying inherently safer design aspects of those programs. This will promote an awareness of the concepts among a broad population of chemists, engineers, and business leaders. This heightened awareness will encourage the application of inherently safer design principles as a part of the normal, everyday work process, instead of something that is only considered and discussed at the safety review meetings.

A number of companies have recently described their inherent safety review practices:

- Bayer (Pilz, 1995) uses a procedure based on hazard analysis, focusing on the application of inherent safety principles to reduce or eliminate hazards.
- Dow (Sheffler, 1996; Gowland, 1996a,b) describes the use of the Dow Fire and Explosion Index (Dow, 1994b) and the Dow Chemical Exposure Index (Dow, 1994a) as measures of inherent safety, along with the use of inherently safer design principles to reduce hazards.
- Exxon Chemical (French et al., 1996; Wixom, 1995) has developed an inherent safety review process, and its application at various points in the process life cycle.
- ICI (Gillett, 1995; Turney, 1990) describes a 6-step hazard review procedure which occurs at specific points in the life cycle of a process, and focuses on inherently safer designs.
- The Rohm and Haas Major Accident Prevention Program (Renshaw, 1990; Berger and Lantzy, 1996; Hendershot, 1991a) is based on potential accident consequence analysis and uses checklists based on inherently safer design principles to identify ways to eliminate or reduce hazards.
- Sandoz (Ankers, 1995) has developed a software tool to assist chemists and engineers in identifying hazards, and inherently safer process options.
- Union Carbide (Lutz, 1995a,b) published a checklist used in their inherent safety reviews.

Inherent Safety Review Objectives

The objectives for an inherent safety review are to employ a synergistic team to:

- Understand the hazards.
- Find ways to reduce or eliminate these hazards.

The first major objective for the inherent safety review is the development of a good understanding of the hazards involved in the process. Early understanding of these hazards provides time for the development team to implement recommendations of the inherent safety effort. Hazards associated with flammability, pressure, and temperature are relatively easy to identify. Reactive chemistry hazards are not. They are frequently difficult to identify and understand in the lab and pilot plant. Special calorimetry equipment and expertise are often necessary to fully characterize the hazards of runaway reactions and decompositions. Similarly, industrial hygiene and toxicology expertise is desirable to help define and understand health hazards associated with the chemicals employed.

Reducing and eliminating hazards and their associated risks is the second major objective. Applying inherent safety principles early in the product/process development effort provides the greatest opportunity to achieve the objectives of the inherent safety review process for the project at hand. If these principles are applied late in the effort the results may have to be applied to the "project after next" as the schedule may not permit implementation of the results.

Experience with inherent safety reviews indicates that project investment costs are frequently reduced as a result of this exercise. Eliminating equipment and reducing the need for safety critical instrumentation are typically the main contributors to investment reduction. Capturing these potential savings depends greatly on the timing of the reviews.

Inherent Safety Review Timing

Inherent safety reviews should be considered during appropriate stages of a typical project's life cycle. Suggested timing includes:

- During the chemistry forming (synthesis) stage for product/process research and development to focus on the chemistry and process (refer to Section 4.2).

- During the facilities design scoping and development prior to completion of the design basis to focus on equipment and configuration (refer to Sections 4.3 and 4.4).
- During regular operations to identify potential improvements for the next plant. Process improvement studies can then demonstrate the feasibility of inherent safety improvements so they can easily be incorporated into the design of the next plant or in major revamps of existing facilities.

Inherent Safety Review Team Composition

The composition of the inherent safety review team will vary depending upon the stages of the development cycle and the nature of the product/process. The team composition is selected from the typical skill areas checked in Table 7.1 and is generally four to seven members. Knowledgeable people, with an appropriate matrix of skills, are required for a successful review effort. The industrial hygienist/toxicologist and chemist are key members of the team to ensure understanding of the hazards associated with reactions, chemicals, intermediates, and products. These members play a key role in explaining hazards, especially where there may be choices among chemicals or processes.

TABLE 7.1

Inherent Safety Review Team Composition

	Product Development	Design Development	Operations
Industrial Hygienist/Toxicologist	✔	✔	✔
Chemist	✔	✔	✔
Process Design	✔	✔	✔
Safety Engineer	✔	✔	✔
Process Technology Leader	✔	✔	✔
Environmental	✔	✔	✔
Control Engineer		✔	✔
Operator			✔
Operations Supervisor		✔	✔
Maintenance			✔

> **I use not only all the brains I have, but all I can borrow.**
> —Woodrow Wilson

Often an organization will strive for the elimination of a specific toxic material from a given process. Alternatives will also have other hazards and risks that require an informed choice. The industrial hygienist, chemist, and safety engineer play an important role in developing the information for making the selection between alternatives.

Inherent Safety Review Process Overview

Good preparation is very important for an effective inherent safety review. Preparation for the review is summarized in Figure 7.1 and includes the following items:

1. Define the desired product.

2. Describe optional routes to manufacture the desired product (if available)

3. Prepare simplified process flow diagram.
   —Include alternative processes.

4. Define chemical reactions.
   —Desired and undesired.
   —Determine potential for runaway reactions/decompositions.

5. List all chemicals and materials employed.
   —Develop a chemical compatibility matrix.
   —Include air, water, rust, etc.

6. Define physical, chemical, and toxic properties.
   —Provide NFPA hazard ratings or equivalent.

7. Define process conditions (pressure, temperature, etc.).

8. Estimate quantities used in each process system (tanks, reactors, etc.).
   —State plant capacity basis.
   —Estimate quantities of wastes/emissions.

9. Define site specific issues such as environmental, regulatory, community, spacing, permitting, etc.

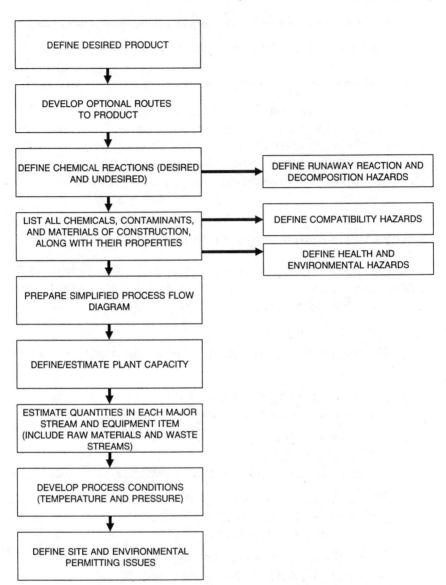

Figure 7.1. Inherent safety review preparation.

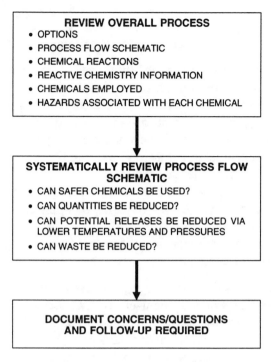

Figure 7.2. Inherent safety review.

After the background information is developed, the inherent safety review can be arranged. The review steps are summarized in Figure 7.2. Steps proposed for the review are as follows:

10. Review the background information from Steps 1 to 9.

11. Define major potential hazards.

12. Systematically review the process flow schematic looking at each process step and hazardous material to identify creative ways to improve the process by applying inherent safety principles to reduce or eliminate hazards.

13. During the inherent safety review, at the design development stage, identify potential human factors/ergonomics issues that should be addressed by the design team.

14. Document the review and follow-up items.

During the systematic review of the process flow schematic (Step 12) the team will examine some of the following questions:

- Can safer chemicals be used? (Nontoxic or nonvolatile reactant.)
- Can quantities be reduced? (Careful look at intermediates storage.)
- Can potential releases be reduced via lower temperatures or pressures, elimination of equipment or by using sealless pumps?
- Can waste be reduced? (Regenerable catalyst or recycle.)
- What additional information is required? (Toxicology information or reactive chemicals data.)
- Can the equipment or process be simplified, be made mistake proof, or at least mistake tolerant, by applying ergonomic/human factors principles?
- Have trace materials or contaminants been identified along with their effects in the process?

It is not unusual to have several inherent safety reviews during the product/process development effort. Early reviews will often not have all of the information required for Steps 1 to 9. The follow-up items will describe what is necessary to obtain the missing information such as toxicity data on new intermediates or products of undesired reactions.

Focus of the Different Inherent Safety Reviews

The focus of the inherent safety reviews will change from the chemistry forming stage through regular operations. While each review is unique, Table 7.2 shows which areas receive emphasis during the different reviews. In the regular operations review, the team is identifying inherent safety opportunities for the next major expansion or grass roots facility. The check marks indicate the relative emphasis each area receives.

In the "chemistry forming" inherent safety review, the team will cover such things as:

- Understanding of the hazards.
- Choice of best route to produce a given chemical or product.
- Process improvement.
  —Reactor types and conditions.
  —Intermediate storage optimization.
  —Waste minimization.
- Requirements for additional information.

TABLE 7.2

Focus of Different Inherent Safety Reviews

	Chemistry and Process Selection	Design Scoping	Regular Operation
Minimize • Reduce quantities	✔	✔✔✔	✔✔
Substitute • Use safer materials	✔✔✔	✔	✔✔
Moderate • Use less hazardous conditions • Reduce wastes	✔✔	✔✔	
Simplify	✔	✔✔✔	✔✔

During design scoping, the team will concentrate on minimizing equipment, reducing inventories, simplifying the process, reducing wastes, and optimizing process conditions.

During process hazards reviews (such as HAZOP), inherent safety concepts are also considered. Mistake proofing the design should receive attention and each safety critical device (last line of defense) and safety critical procedure should be examined to see if there is a way to eliminate the need for the device or procedure.

When the inherent safety process has been expanded to review regular or routine operation, the team will look at all aspects of inherent safety to provide suggested improvements for both the existing facility and for the next plant.

Inherent Safety Review Initiates Project Human Factors/Ergonomics Effort

As noted in Step 13 of the inherent safety review at the design development stage, the team reviews the process to develop a list of potential ergonomics/human factors issues. This list provides input to the design team so that these issues can be addressed as the design progresses.

Operations and human factors specialist involvement is important to find optimum solutions to human factors/ergonomics issues. Applying inherent safety principles to ergonomics/human factors issues can reduce risks associated with:

- Materials handling of bags, drums, solids, etc.
- Tankcar, tankwagon, and hopper car loading, unloading, and cleaning.
- Manual operations such as filter cleaning, sampling, batch operations.
- Hose handling at manifolds, docks etc.
- Safety critical procedures.
- Equipment isolation for maintenance.

Good Preparation Required for Effective Inherent Safety Reviews

The completeness of the information described in Steps 1 to 9 prior to the review will determine the quality of the inherent safety review. The chemist needs to define the desired reactions, and to develop an understanding of potential side reactions. Effects on reaction chemistry need to be developed for mischarges or process deviations. These information requirements on process chemistry are discussed in Section 4.2.

Reaction kinetics are often difficult to develop in the normal experimental laboratory or pilot plant. Specialized reaction calorimeters and reactive chemistry testing (differential scanning calorimeter, DSC; Accelerating Rate Calorimeter, ARC™; vent sizing package, VSP; etc.) are required to obtain thermodynamic and kinetic information (CCPS, 1995d). Many chemical companies have established reactive chemistry laboratories with reaction engineering specialists to assist in developing this information. Consulting firms also provide this expertise and experimental capability.

Industrial hygiene/toxicology people need to develop acute and chronic toxicology information on all the materials used and produced in the process. This information should also include the potential products of abnormal reactions. The industrial hygiene member of the team should be prepared to explain the toxicology information on the material safety data sheet (MSDS) to the review team.

Physical property data on the process chemicals are necessary. This information includes such items as melting point, boiling point, vapor pressure, water solubility, flammability data, and odor threshold, to name a few.

This information must be organized and coordinated to satisfy a company's process and facilities documentation process and regulatory requirements.

Reactive Chemicals Screening

Experience has shown that reactive chemistry hazards are sometimes undetected during bench scale and pilot plant development of new products and processes. Reactive chemistry hazards must be identified so they can be addressed in the inherent safety review process. Chemists should be encouraged and trained to explore reactive chemistry of "off-normal" operations. Simple reactive chemicals screening tools, such as the interactions matrix described in Section 4.2, can be used by R&D chemists.

Time Required for an Inherent Safety Review

Low hazard process reviews can be completed in several hours. Reviews of complex projects are normally completed in one to two days. The key to an effective and efficient review is good preparation.

7.2. Inherent Safety Review Training

Training requirements for inherent safety begin with the need for general awareness for engineers, scientists, and business leaders. This awareness training should include the concepts of inherent safety along with how inherent safety is implemented in the organization's process safety management, new product development, and project execution programs. More specific training should be considered for those people who play key roles in inherently safer processes such as:

- R&D chemists involved in product and process development.
- Process development and process design engineers involved in process scoping, development, and design activities.
- Functional specialists such as safety engineers and industrial hygienists.

If an organization chooses to include inherent safety reviews in its implementation approach, training should be included on how to conduct an inherent safety review.

The Institution of Chemical Engineers has recently updated the "Inherently Safer Process Design" safety training package (IChemE and IPSG, 1995). Included in this package are course materials, videos, a good variety of case studies, and guidance for the course leader. This is an excellent course which can be presented by itself or used to augment inherent safety training courses designed to meet an organization's specific needs.

Companies may wish to develop workshops to train potential team members in the inherent safety review process. The workshop can provide background information on inherent safety concepts, the extensive systems required to manage hazardous materials, and information on the inherent safety review process. Videos, problems, examples, and team exercises can be included to enliven the education process.

7.3. Summary

Implementing an inherent safety review process is one mechanism companies can use to institutionalize inherent safety. The review process should integrate well with company systems for process safety management, new product development, and project execution. Safety, health, and environmental considerations in the new product or process development effort can be strengthened via the introduction of the inherent safety review. Companies may also build inherently safer design concepts into their existing process safety management system and process hazard reviews.

8

Future Initiatives

Process safety will be improved by applying inherently safer technology to new and existing processes. The greatest gains in inherent safety will come from the invention of new chemistry and new processing technologies which eliminate the need for hazardous materials and operations. A key to the future invention of this new technology is establishing a broad awareness of inherently safer technology within each company and among the entire research and chemical engineering community, including academia. This awareness is especially important among chemists and others working on the earliest development of process chemistry. Initiatives such as those presented below would lead to a quicker application of this technology.

8.1. Dissemination of Inherently Safer Technology

Incorporating Inherently Safer Technology into Process Safety Management

Inherently safer technology within each company should be incorporated into process safety management programs to make its application a "way of life" and therefore encourage invention of inherently safer processes and procedures. The best way to incorporate inherently safer technology in process safety management programs must be identified. Should inherently safer technology be separate and distinct from process hazards analysis programs, or should it be an integral part of the process hazards analysis programs? What are the advantages and disadvantages of each method of application?

Incident investigations are a part of process safety management. In investigations, lessons are learned as to how inherently safer technology could have prevented or mitigated the results. How can these learnings be disseminated such that future incidents in similar processes are avoided?

There are barriers impeding progress in applying inherently safer technology. How might they be best identified and overcome?

Encouraging Invention within the Chemical and Chemical Engineering Community

Publicizing the virtues of inherently safer technology beyond the process safety community and into the broader chemistry and chemical engineering community is necessary. The authors of this book encourage readers to look for opportunities to "beat the drum" for inherently safer technology at every opportunity as they interact with this more broad audience. Awareness can be raised in a variety of ways. Books by Trevor Kletz (1984, 1991b), by CCPS (1993a), and by Englund (1993) have proven to be successful vehicles. The IChemE and IPSG Inherently Safer Processing training program (IChemE and IPSG, 1995) is another successful format for promoting this topic. Other means need to be explored and exploited.

Including Inherently Safer Technology into the Education of Chemists and Chemical Engineers

Teaching inherently safer technology concepts in undergraduate chemistry, chemical engineering, and related disciplines will be a great benefit as students move into industry after graduation.

8.2. Developing Inherently Safer Technology Databases and Libraries

Readily available inherently safer technology databases which are cataloged, cross referenced and indexed need to be developed. These might take the form of libraries of information. Several examples of needed databases are:

- A continually updated database including submitted descriptions of inherently safer technology successes and failures.
- A collection of databases of chemicals and of functional groups which rank chemicals and groups relative to their reactivity, stability, toxicity, and flammability categories. This would assist in the evaluation of the potential benefits of substituting one, somewhat safer, chemical for another.
- A database of the hazards associated with different types of equipment and unit operations including the applicability of inherently safer technology in each. As innovative solutions to hazards in equipment and process operations are discovered these could be included in this database for use by others in reducing risk in similar equipment and processes. A summary of design approaches for a number of common types of chemical process equipment will be published in CCPS (1997). This summary may be a starting point for the development of this database.

8.3. Developing Tools to Apply Inherently Safer Technology

The Broad View and Life Cycle Cost of Alternatives

One of the recognized barriers to the successful application of inherently safer technology is the lack of appreciation of the benefits that can be derived from viewing a process broadly rather than narrowly. Employing a cradle-to-grave and a feed-end-to-product-end view will lead to the development of processes which are as inherently safe as possible. Several examples of myopic design are:

- A chemist may think only of optimizing the route of synthesis to avoid a runaway reaction hazard and not consider the safety and design implications of flammable reaction by-products.

- A designer fails to identify and consider the environmental land use implications of dismantling a process at the end of its useful life.

At any given time individuals are more concerned with a particular unit operation, or with a particular life cycle stage of the process. A more broad approach, however, would require an analysis of the impact of the process on:

- upstream and downstream operations
- ancillary operations such as waste stream management and the handling of byproducts
- later stages in the process life cycle

Tools to apply such a more broad view of a process would pay inherently safer dividends. One tool might include instructions on how to estimate the life cycle cost of proposed alternative solutions. Such a tool is not presently fully developed and available in the public domain. Training to assist in estimating life cycle cost is needed.

Benefits of Reliability Analysis

The reliability of safety interlock systems depends not only on correct installation configuration, but also on the use of quality hardware and on prudent inspection and test programs. In this way the frequency of failure and the duration of failure are minimized, thus reducing the likelihood of such systems being successfully challenged by real events. Guidance in assessing such reliability could lead to simplifying inter-lock systems (Green and Dowell, 1995; 1996).

Direction is necessary to analyze the reliability of overall processes and the individual unit operations within processes. High process reliability may minimize the necessity of large in-process inventories by minimizing the need for storage facilities of process intermediates and minimizing the need for storage and blending facilities for "off-spec" product. It may also maximize the ability to produce product "as needed," thereby reducing on-plant product storage requirements. Highly reliable management of raw material purchasing and delivery systems may minimize the need for on-plant raw material storage.

Potential Energy

It is obvious that, all other things being equal, a process operated at high temperature and high pressure, containing exothermically reactive materials, and containing a large inventory is inherently less safe than one at ambient temperature and pressure, containing thermally stable materials, and containing a small inventory. A measure of the magnitude of this difference in safety is the difference in the potential energy in the process. By comparing the potential energies of alternative process strategies, the lowest energy alternative can be identified. Methods and guidance to estimate the total potential energy in different types of situations encountered in processes are needed.

A Table of Distances

Appropriate spacing of unit operations within a process and appropriate spacing of a process from other processes, from employees nonessential to day-to-day process operation, and from the public is inherently safer. A definition of "appropriate spacing" would assist in evaluating the process location alternatives. This definition may take the form of a table of distances as a function of the type of hazard, inventory quantity and other factors.

Quantitative Measures of Inherently Safer Technology

Several available quantitative risk indices can be used to measure the degree of application of inherently safer technology in a process. Examples include the Dow Chemical Exposure Index (Dow, 1994a), the Dow Fire and Explosion Index (Dow, 1994b), the Mond Index (ICI, 1985; Tyler, 1985), and the Prototype Index of Inherent Safety by Edwards et al. (1996). Refinement of these quantitative measurement techniques and convergence to a single set of accepted indices would be beneficial. Accepted metrics would aid in the comparison of alternatives and in quantifying the degree of process improvement that the various applications of inherently safer technology bring.

Appendix

A Sample Inherently Safer Process Checklist

The following checklist contains a number of questions which can aid in identifying inherently safer process options. The list is adapted from CCPS (1993a). Other checklists, particularly the extensive checklist in Appendix B of the *Guidelines for Hazard Evaluation Procedures, 2nd Edition with Worked Examples* (CCPS, 1992) contain many questions which are related to inherent safety.

1. Minimize

❑ Have all in-process inventories of hazardous materials in storage tanks been minimized?

❑ Are all of the proposed in-process storage tanks really needed?

❑ Has all processing equipment handling hazardous materials been designed to minimize inventory?

❑ Is process equipment located to minimize length of hazardous material piping?

❑ Can piping sizes be reduced to minimize inventory?

❑ Can other types of unit operations or equipment reduce material inventories? For example:
   —Wiped film stills in place of continuous still pots
   —Centrifugal extractors in place of extraction columns
   —Flash dryers in place of tray dryers
   —Continuous reactors in place of batch
   —Plug flow reactors in place of CFSTRs
   —Continuous in-line mixers in place of mixing vessels

❑ Is it possible to feed hazardous materials (for example, chlorine) as a gas instead of liquid, to reduce pipeline inventories?

❑ Is it possible to generate hazardous reactants "in-situ" from less hazardous raw materials?

❑ Is it possible to generate hazardous reactants on site from less hazardous materials, minimizing the need to store or transport large quantities of hazardous materials?

2. Substitution/Elimination

❑ Is it possible to completely eliminate hazardous raw materials, process intermediates, or by-products by using an alternative process or chemistry?

❑ Is it possible to completely eliminate in-process solvents by changing chemistry or processing conditions?

❑ Is it possible to substitute less hazardous raw materials?
   —Noncombustible rather than flammable solvents
   —Less volatile raw materials
   —Less toxic raw materials
   —Less reactive raw materials
   —More stable raw materials

❑ Is it possible to substitute less hazardous final product solvents?

❑ For equipment containing materials which become unstable at elevated temperature or freeze at low temperature, is it possible to use heating and cooling media which limit the maximum and minimum temperatures attainable?

3. Moderate

❑ Can the supply pressure of raw materials be limited to less than the working pressure of the vessels they are delivered to?

❑ Can reaction conditions (temperature, pressure) be made less severe by using a catalyst, or by using a better catalyst?

❑ Can the process be operated at less severe conditions? If this results in lower yield or conversion, can raw material recycle compensate for this loss?

❑ Is it possible to dilute hazardous raw materials to reduce the hazard potential? For example:
—Aqueous ammonia instead of anhydrous
—Aqueous HCl instead of anhydrous
—Sulfuric acid instead of oleum
—Dilute nitric acid instead of concentrated fuming nitric acid
—Wet benzoyl peroxide instead of dry

4. Simplify

❑ Can equipment be designed sufficiently strong to totally contain the maximum pressure generated, even if the "worst credible event" occurs?

❑ Is all equipment designed to totally contain the materials which might be present inside at ambient temperature or the maximum attainable process temperature (i.e., don't rely on the proper functioning of external systems such as refrigeration systems to control temperature such that vapor pressure is less than equipment design pressure)?

❑ Can several process steps be carried out in separate processing vessels rather than a single multipurpose vessel? This reduces complexity and the number of raw materials, utilities, and auxiliary equipment connected to a specific vessel, thereby reducing the potential for hazardous interactions.

❑ Can equipment be designed such that it is difficult or impossible to create a potential hazardous situation due to an operating error (for example, by opening an improper combination of valves)?

5. Location/Siting/Transportation

☐ Can process units be located to reduce or eliminate adverse impacts from other adjacent hazardous installations?

☐ Can process units be located to eliminate or minimize:
   Off-site impacts?
—Impacts to employees on-site?
—Impacts on other process or plant facilities?

☐ Can the plant site be chosen to minimize the need for transportation of hazardous materials and to use safer transport methods and routes?

☐ Can a multistep process, where the steps are done at separate sites, be divided up differently to eliminate the need to transport hazardous materials?

References

9.1. Chapter References

Agreda, V. H., L. R. Partin, and W. H. Heise (1990). "High-Purity Methyl Acetate Via Reactive Distillation." *Chemical Engineering Progress* (February), 40–46.

Akay, G., and B. J. Azzopardi (1995). *Proceedings of the First International Conference on Science, Engineering and Technology of Intensive Processing*, September 18–20, 1995, Nottingham, U. K., Nottingham, U. K.: University of Nottingham.

Allen, D. (1992). "The Role of Catalysis in Industrial Waste Reduction." *Industrial Environmental Chemistry*, ed. D. T. Sawyer, and A. E. Martell, 89–98. New York: Plenum Press.

Althaus, V. E., and S. Mahalingam (1992). "Inherently Safer Process Designs." *South Texas Section AIChE Process Plant Safety Symposium*, February 18–19, 1992, Houston, TX, ed. W. F. Early, V. H. Edwards, and E. A. Waltz, 546–555. Houston, TX: American Institute of Chemical Engineers South Texas Section.

The American College Dictionary (1967). New York: Random House.

American Institute of Chemical Engineers (AIChE) (1995). *CHEMPAT: A Program to Assist Hazard Evaluation and Management*. New York: American Institute of Chemical Engineers (Publication Z-1).

American Society for Testing and Materials (ASTM) (1994). *CHETAH, Version 7.0: The ASTM Computer Program for Chemical Thermodynamic and Energy Release Evaluation*. ASTM Data Series DS 51B. Philadelphia, PA: American Society for Testing and Materials.

Ankers, R. (1995). "Introducing Inherently Safer Concepts Early in Process Development With PRORA." *CCPS Inherently Safer Process Workshop*, May 17, 1995, Chicago, IL.

Ashford, N. A. (1993). *The Encouragement of Technological Change for Preventing Chemical Accidents: Moving Firms From Secondary Prevention and Mitigation to Primary Prevention*. Cambridge, MA: Center for Technology, Policy and Industrial Development, Massachusetts Institute of Technology.

Benson, R. S., and J. W. Ponton (1993). "Process Miniaturisation—A Route to Total Environmental Acceptability?" *Trans. IChemE* 71, Part A (March), 160–168.

Berger, S. A., and R. J. Lantzy (1996). "Reducing Inherent Risk Through Consequence Modeling." *1996 Process Plant Safety Symposium*, Volume 1, April 1–2, 1996, Houston, TX, ed. H. Cullingford, 15–23. Houston, TX: South Texas Section of the American Institute of Chemical Engineers.

Bodor, N. (1995). "Design of Biologically Safer Chemicals." *Chemtech* (October), 22–32.

Bretherick, L. (1995). *Handbook of Reactive Chemical Hazards*. 5th Edition, 2 Volumes, Oxford, U.K.: Butterworth-Heinemann.

Burch, W. M. (1986). "Process Modifications and New Chemicals." *Chemical Engineering Progress* (April), 5–8.

Carrithers, G. W., A. M. Dowell, and D. C. Hendershot (1996). "It's Never Too Late for Inherent Safety." *International Conference and Workshop on Process Safety Management and Inherently Safer Processes*, October 8–11, 1996, Orlando, FL, New York: American Institute of Chemical Engineers.

Catanach, J. S., and S. W. Hampton (1992). "Solvent and Surfactant Influence on Flash Points of Pesticide Formulations." *ASTM Spec. Tech. Publ.* 11, 149–57.

Center for Chemical Process Safety (CCPS) (1988a). *Guidelines for Safe Storage and Handling of High Toxic Hazard Materials*. New York: American Institute of Chemical Engineers.

Center for Chemical Process Safety (CCPS) (1988b). *Guidelines for Vapor Release Mitigation*. New York: American Institute of Chemical Engineers.

Center for Chemical Process Safety (CCPS) (1989a). *Guidelines for Chemical Process Quantitative Risk Analysis*. New York: American Institute of Chemical Engineers.

Center for Chemical Process Safety (CCPS) (1989b). *Guidelines for Technical Management of Chemical Process Safety*. New York: American Institute of Chemical Engineers.

Center for Chemical Process Safety (CCPS) (1992). *Guidelines for Hazard Evaluation Procedures, Second Edition With Worked Examples*. New York: American Institute of Chemical Engineers.

Center for Chemical Process Safety (CCPS) (1993a). *Guidelines for Engineering Design for Process Safety*. New York: American Institute of Chemical Engineers.

Center for Chemical Process Safety (CCPS) (1993b). *Guidelines for Safe Automation of Chemical Processes*. New York: American Institute of Chemical Engineers.

Center for Chemical Process Safety (CCPS) (1993c). *Guidelines for Auditing Process Safety Management Systems*. New York: American Institute of Chemical Engineers.

Center for Chemical Process Safety (CCPS) (1994a). *Guidelines for Preventing Human Error in Process Safety*. New York: American Institute of Chemical Engineers.

Center for Chemical Process Safety (CCPS) (1994b). *Guidelines for Evaluating the Characteristics of Vapor Cloud Explosions, Flash Fires, and BLEVES*. New York: American Institute of Chemical Engineers.

Center for Chemical Process Safety (CCPS) (1995a). *Tools for Making Acute Risk Decisions With Chemical Process Safety Applications*. New York: American Institute of Chemical Engineers.

Center for Chemical Process Safety (CCPS) (1995b). *Guidelines for Chemical Transportation Risk Analysis*. New York: American Institute of Chemical Engineers.

Center for Chemical Process Safety (CCPS) (1995c). *Guidelines for Technical Planning for On-Site Emergencies*. New York: American Institute of Chemical Engineers.

Center for Chemical Process Safety (CCPS) (1995d). *Guidelines for Chemical Reactivity Evaluation and Application to Process Design*. New York: American Institute of Chemical Engineers.

Center for Chemical Process Safety (CCPS) (1995e). *Guidelines for Safe Storage and Handling of Reactive Materials*. New York: American Institute of Chemical Engineers.

Center for Chemical Process Safety (CCPS) (1996). *Guidelines for Writing Effective Operating and Maintenance Procedures*. New York: American Institute of Chemical Engineers.

Center for Chemical Process Safety (CCPS) (1997). [scheduled publication]. *Guidelines for Selecting the Design Bases of Process Safety Systems*. New York: American Institute of Chemical Engineers.

Collins, R. L. (1978). *Flying IFR*. New York: Delacorte Press/E. Friede.

Crowl, D. A., and J. F. Louvar (1990). *Chemical Process Safety Fundamentals With Applications*, 14–15. Englewood Cliffs, NJ: Prentice Hall.

Dale, S. E. (1987). "Cost Effective Design Considerations for Safer Chemical Plants." *Proceedings of the International Symposium on Preventing Major Chemical Accidents*, February 3–5, 1987, Washington, D. C., ed. J. L. Woodward, 3.79–3.99. New York: American Institute of Chemical Engineers.

Dartt, C. B., and M. E. Davis (1994). "Catalysis for Environmentally Benign Processing." *Ind. Eng. Chem. Res.* 33, 2887–99.

Davis, G. A., L. Kincaid, D. Menke, B. Griffith, S. Jones, K. Brown, and M. Goergen (1994). *The Product Side of Pollution Prevention: Evaluating the Potential for Safe Substitutes*. Cincinnati, Ohio: Risk Reduction Engineering Laboratory, Office of Research and Development, U. S. Environmental Protection Agency.

The Design of Inherently Safer Plants (1988). *Chemical Engineering Progress* (September), 21.

DeSimone, J. M., E. E. Maury, Z. Guan, J. R. Combes, Y. Z. Menceloglu, M. R. Clark, J. B. McClain, T. J. Romack, and C. D. Mistele (1994). "Homogeneous and Heterogeneous Polymerizations in Environmentally Responsible Carbon Dioxide." *Preprints of Papers Presented at the 208th ACS National Meeting*, August 21–25, 1994, Washington, DC, 212–214. Center for Great Lakes Studies, University of Wisconsin–Milwaukee, Milwaukee, WI: Division of Environmental Chemistry, American Chemical Society.

Doherty, M., and G. Buzad (1992). "Reactive Distillation by Design." *The Chemical Engineer* (27 August), s17–s19.

Dow Chemical Company (1994a). *Dow's Chemical Exposure Index Guide*. 1st Edition. New York: American Institute of Chemical Engineers.

Dow Chemical Company (1994b). *Dow's Fire and Explosion Index Hazard Classification Guide*. 7th Edition. New York: American Institute of Chemical Engineers.

Edwards, D. W., D. Lawrence, and A. G. Rushton (1996). "Quantifying the Inherent Safety of Chemical Process Routes." 5th World Congress of Chemical Engineering, July 14–18, 1996, San Diego, CA, Paper 52d. New York: American Institute of Chemical Engineers.

Englund, S. M. (1990). "Opportunities in the Design of Inherently Safer Chemical Plants." *Advances in Chemical Engineering* 15, 69–135.

Englund, S. M. (1991a). "Design and Operate Plants for Inherent Safety—Part 1." *Chemical Engineering Progress* (March), 85–91.

Englund, S. M. (1991b). "Design and Operate Plants for Inherent Safety—Part 2." *Chemical Engineering Progress* (May), 79–86.

Englund, S. M. (1993). "Process and Design Options for Inherently Safer Plants." *Prevention and Control of Accidental Releases of Hazardous Gases*, ed. V. M. Fthenakis, 9–62. New York: Van Nostrand Reinhold.

Federal Emergency Management Agency (FEMA), U. S. Department of Transportation (DOT), and U. S. Environmental Protection Agency (EPA) (ca. 1989). *Handbook of Chemical Hazard Analysis Procedures*. Washington, D. C.: FEMA Publications Office.

Flam, F. (1994). "Laser Chemistry: The Light Choice." *Science* 266 (14 October), 215–17.

Forsberg, C. W., D. L. Moses, E. B. Lewis, R. Gibson, R. Pearson, W. J. Reich, G. A. Murphy, R. H. Staunton, and W. E. Kohn (1989). *Proposed and Existing Passive and Inherent Safety-Related Structures, Systems, and Components (Building Blocks) for Advanced Light Water Reactors*. Oak Ridge, TN: Oak Ridge National Laboratory.

Frank, W. L. (1995). "Evaluation of a Containment Building for a Liquid Chlorine Unloading Facility." *Proceedings of the 29th Annual Loss Prevention Symposium*, July 30–August 2, 1995, Boston, MA, ed. E. D. Wixom and R. P. Benedetti, Paper 5b. New York: American Institute of Chemical Engineers.

Freeman, R. A. (1996). Personal Communication to D. C. Hendershot, June 6, 1996.

French, R. W., D. D. Williams, and E. D. Wixom (1996). "Inherent Safety, Health and Environmental (SHE) Reviews." *Process Safety Progress* 15, 1 (Spring), 48–51.

Gay, D. M., and D. J. Leggett (1993). "Enhancing Thermal Hazard Analysis Awareness With Compatibility Charts." *Journal of Testing and Evaluation* 21, 6, 477–80.

Gillett, J. (1995). "Validation Hazards." *The Chemical Engineer* (14 September), 26–28.

Goldschmidt, G., and P. Filskov (1990). "Substitution—A Way to Obtain Protection Against Harmful Substances at Work." *Staub–Reinhaltund Der Luft* 50, 403–5.

Govardhan, C. P., and A. L. Margolin (1995). "Extremozymes for Industry—From Nature and By Design." *Chemistry & Industry* (4 September), 689–93.

Gowland, R. T. (1996a). "Applying Inherently Safer Concepts to a Phosgene Plant Acquisition." *Process Safety Progress* 15, 1 (Spring), 52–57.

Gowland, R. T. (1996b). "Putting Numbers on Inherent Safety." *Chemical Engineering* 103, 3 (March), 82–86.

Green, D. L., and A. M. Dowell (1995). "How to Design, Verify, and Validate Emergency Shutdown Systems." *ISA Transactions* 34, 261–72.

Green, D. L., and A. M. Dowell (1996). "Cookbook Safety Shutdown System Design." *1996 Process Plant Safety Symposium*, Volume 1, April 1–2, 1996, Houston, TX, ed. H. Cullingford, 552–565. Houston, TX: South Texas Section of the American Institute of Chemical Engineers.

Hall, N. (1994). "Chemists Clean Up Synthesis With One-Pot Reactions." *Science* 266 (7 October), 32–34.

Harris, N. C. (1987). "Mitigation of Accidental Toxic Gas Releases." *Proceedings of the International Symposium on Preventing Major Chemical Accidents*, February 3–5, 1987, Washington, D. C., ed. J. L. Woodward, 3.139–3.177. New York: American Institute of Chemical Engineers.

Hendershot, D. C. (1987). "Safety Considerations in the Design of Batch Processing Plants." *Proceedings of the International Symposium on Preventing Major Chemical Accidents*, February 3–5, 1987, Washington, D. C., ed. J. L. Woodward, 3.2–3.16. New York: American Institute of Chemical Engineers.

Hendershot, D. C. (1991a). "Design of Inherently Safer Chemical Processing Facilities." *Texas Chemical Council Safety Seminar*, June 11, 1991, Galveston, TX, Session D.

Hendershot, D. C. (1991b). "The Use of Quantitative Risk Assessment in the Continuing Risk Management of a Chlorine Handling Facility." *The Analysis, Communication, and Perception of Risk*, ed. B. J. Garrick, and W. C. Gekler, 555–65. New York: Plenum Press.

Hendershot, D. C. (1995a). "Conflicts and Decisions in the Search for Inherently Safer Process Options." *Process Safety Progress* 14, 1 (January), 52–56.

Hendershot, D. C. (1995b). "Some Thoughts on the Difference Between Inherent Safety and Safety." *Process Safety Progress* 14, 4 (October), 227–28.

Hendershot, D. C. (1996). "Risk Guidelines As a Risk Management Tool." *1996 Process Plant Safety Symposium*, April 1–2, 1996, Houston, TX, Houston, TX: South Texas Section of the American Institute of Chemical Engineers.

Hochheiser, S. (1986). *Rohm and Haas: History of a Chemical Company*. Philadelphia, PA: University of Pennsylvania Press.

Imperial Chemical Industries (ICI) (1985). *The Mond Index, Second Edition*. Winnington, Northwich, Chesire, U. K.: Imperial Chemical Industries PLC.

The Institution of Chemical Engineers (IChemE), and The International Process Safety Group (IPSG) (1995). *Inherently Safer Process Design*. Rugby, England: The Institution of Chemical Engineers.

Instrument Society of America (ISA) (1996). ISA-S84.01. *Application of Safety Instrumented Systems for the Process Industries*. Research Triangle Park, NC: Instrument Society of America.

ISO 9000 (1994). *Quality Management and Quality Assurance Standards—Guidelines for Selection and Use*. Geneva, Switzerland: International Organization for Standardization.

Johnston, K. P. (1994). "Safer Solutions for Chemists." *Nature* 368 (17 March), 187–88.

Kepner, C. H., and B. B. Tregoe (1981). *The New Rational Manager*. Princeton, NJ: Princeton Research Press.

Kharbanda, O. P., and E. A. Stallworthy (1988). *Safety in the Chemical Industry*. London: Heinemann Professional Publishing, Ltd.

Kletz, T. A. (1978). "What You Don't Have, Can't Leak." *Chemistry and Industry* (6 May), 287-92.

Kletz, T. A. (1984). *Cheaper, Safer Plants, or Wealth and Safety at Work*. Rugby, Warwickshire, England: The Institution of Chemical Engineers.

Kletz, T. A. (1985). "Inherently Safer Plants." *Plant/Operations Progress* 4, 3 (July), 164–67.

Kletz, T. A. (1988). *Learning from Accidents*. Oxford, UK: Butterworth-Heinemann.

Kletz, T. A. (ca. 1988). Seminar Presentation. Union Carbide Corporation.

Kletz, T. A. (1991a). "Billiard Balls and Polo Mints." *The Chemical Engineer*, 495 (25 April), 21–22.

Kletz, T. A. (1991b). *Plant Design for Safety*. New York: Hemisphere.

Kletz, T. A. (1996). "Inherently Safer Design—The Growth of an Idea." *Process Safety Progress* 15, 1 (Spring), 5–8.

Lewis, B., and G. von Elbe (1987). *Combustion Flames and Explosions of Gases*. 3rd Edition. Orlando, FL: Academic Press.

Lewis, D. J. (1979). "The Mond Fire, Explosion and Toxicity Index Applied to Plant Layout and Spacing." *13th Annual Loss Prevention Symposium*, April 2–5, 1979, Houston, TX, 20–26. *Loss Prevention*, No. 13. New York: American Institute of Chemical Engineers.

Lin, D., A. Mittelman, V. Halpin, and D. Cannon (1994). *Inherently Safer Chemistry: A Guide to Current Industrial Processes to Address High Risk Chemicals*. Washington, DC: Office of Pollution Prevention and Toxics, U. S. Environmental Protection Agency.

Lorenzo, D. K. (1990). *A Manager's Guide to Reducing Human Errors: Improving Human Performance in the Chemical Industry*. Washington, D. C.: Chemical Manufacturers Association.

Lutz, W. K. (1995a). "Take Chemistry and Physics into Consideration in All Phases of Chemical Plant Design." *Process Safety Progress* 14, 3 (July), 153–62.

Lutz, W. K. (1995b). "Putting Safety into Chemical Plant Design." *Chemical Health and Safety* 2, 6 (November/December), 12–15.

Mandich, N. V., and G. A. Krulik (1992). "Substitution of Nonhazardous for Hazardous Process Chemicals in the Printed Circuit Industry." *Me. Finish.* 90, 11, 49–51.

Mansfield, D. P. (1994). *Inherently Safer Approaches to Plant Design*. Warrington, Cheshire, U. K.: United Kingdom Atomic Energy Authority.

Mansfield, D. P. (1996). "Viewpoints on Implementing Inherent Safety." *Chemical Engineering* 103, 3 (March), 78–80.

Manzer, L. E. (1993). "Toward Catalysis in the 21st Century Chemical Industry." *Catalysis Today* 18, 199–207.

Manzer, L. E. (1994). "Chemistry and Catalysis." *Benign by Design: Alternative Synthetic Design for Pollution Prevention*, ed. P. T. Anastas, and C. A. Farris, 144–54. Washington, D. C.: American Chemical Society.

Marshall, J., A. Mundt, M. Hult, T. C. McKealvy, P. Myers, and J. Sawyer (1995). "The Relative Risk of Pressurized and Refrigerated Storage for Six Chemicals." *Process Safety Progress* 14, 3 (July), 200–211.

Marshall, V. C. (1990). "The Social Acceptability of the Chemical and Process Industries." *Trans. IChemE* 68, Part B (May), 83–93.

Marshall, V. C. (1992). "The Management of Hazard and Risk." *Applied Energy* 42, 63–85.

McQuaid, J. (1991). "Know Your Enemy: The Science of Accident Prevention." *Trans. IChemE* 69, Part B (February), 9–19.

Medard, L. A. (1989). *Accidental Explosions*. 2 Volumes, West Sussex, England: Ellis Horwood Limited.

Misono, M., and T. Okuhara (1993). "Solid Superacid Catalysts." *Chemtech*, November, 23–29.

Mizerek, P. (1996). "Disinfection Techniques for Water and Wastewater." *The National Environmental Journal* (January/February), 22–28.

National Fire Protection Association (NFPA) (1992). *Explosion Prevention Systems*. NFPA 69, Quincy, MA: NFPA.

National Fire Protection Association (NFPA) (1994). *Fire Hazard Properties of Flammable Liquids, Gases, and Volatile Solids*. NFPA 325, Quincy, MA: NFPA.

Negron, R. M. (1994). "Using Ultraviolet Disinfection in Place of Chlorination." *The National Environmental Journal* (March/April), 48–50.

Norman, D. A. (1988). *The Psychology of Everyday Things*. New York: Basic Books.

Norman, D. A. (1992). *Turn Signals Are the Facial Expressions of Automobiles*. Reading, MA: Addison-Wesley Publishing Company.

Occupational Safety and Health Administration (OSHA) (1992). 29 CFR Part 1910. "Process Safety Management of Highly Hazardous Chemicals; Explosives and Blasting Agents; Final Rule." *Federal Register* 57, 36 (February 24), 6356–417.

Paint Removers: New Products Eliminate Old Hazards (1991). *Consumer Reports* (May), 340-343.

Parshall, D. R. (1989). "Suggestions for Structuring the Product Design and Manufacturing Process to Help Create Safe Products and Reduce Litigation Risk." *Hazard Prevention* (April/June), 12–17.

Patty's Industrial Hygiene and Toxicology (1993). Fourth Edition, ed., G. D. Clayton and F. E. Clayton. New York: John Wiley and Sons.

Petroski, H. (1985). *To Engineer Is Human: The Role of Failure in Successful Design*. New York: St. Martin's Press (reprinted in 1992 by Vintage Books, New York).

Petroski, H. (1995). *Engineers of Dreams: Great Bridge Builders and the Spanning of America*. New York: Alfred A. Knopf.

Pilz, V. (1995). "Bayer's Procedure for the Design and Operation of Safe Chemical Plants." *Inherently Safer Process Design*, 4.54–4.65. Rugby, England: The Institution of Chemical Engineers.

Ponton, J. (1996). "Some Thoughts on the Batch Plant of the Future." *5th World Congress of Chemical Engineering*, July 14–18, 1996, San Diego, CA, Paper 52e. New York: American Institute of Chemical Engineers.

Puglionesi, P. S., and R. A. Craig (1991). "State-of-the-Art Techniques for Chlorine Supply Release Prevention." *Environmental Analysis, Audits and Assessments—Papers From the 84th Annual Meeting and Exhibition of the Air and Waste Management Association*, June 16–21, 1991, Vancouver, British Columbia, Canada, 91-145.5. Pittsburgh, PA: Air and Waste Management Association.

Puranik, S. A., K. K. Hathi, and R. Sengupta (1990). "Prevention of Hazards Through Technological Alternatives." *Safety and Loss Prevention in the Chemical and Oil Processing Industries*, October 23–27, 1989, Singapore, 581–587. IChemE Symposium Series, No. 120. Rugby, Warwickshire, U. K.: The Institution of Chemical Engineers.

Purdy, G., and M. Wasilewski (1995). "Focused Risk Management for Chlorine Installations." *Mod. Chlor-Alkali Technol.* 6, 32–47.

Raghaven, K. V. (1992). "Temperature Runaway in Fixed Bed Reactors: Online and Offline Checks for Intrinsic Safety." *Journal of Loss Prevention in the Process Industries* 5, 3, 153–59.

Rand, G. M., and S. R. Petrocelli (1985). *Fundamentals of Aquatic Toxicology*. New York: Hemisphere.

Reid, R. A., and D. C. Christensen (1994). "Evaluate Decision Criteria Systematically." *Chemical Engineering Progress* (July), 44–49.

Renshaw, F. M. (1990). "A Major Accident Prevention Program." *Plant/Operations Progress* 9, 3 (July), 194–97.

Rogers, R. L., and S. Hallam (1991). "A Chemical Approach to Inherent Safety." *IChemE Symposium Series* No. 124, 235–41.

Rogers, R. L., D. P. Mansfield, Y. Malmen, R. D. Turney, and M. Verwoerd (1995). "The INSIDE Project: Integrating Inherent Safety in Chemical Process Development and Plant Design." *International Symposium on Runaway Reactions and Pressure Relief Design*, August 2-4, 1995, Boston, MA, ed. G. A. Melhem and H. G. Fisher, 668–689. New York: American Institute of Chemical Engineers.

Rothenberg, D. H., and I. Nimmo (1996). "The Concept of Abnormal Situation Management and Mechanical Reliability." *1996 Process Plant Safety Symposium*, April 1–2, 1996, Houston, Texas. Volume 2, ed. H. Cullingford, pages 193–208. Houston, TX: South Texas Section of the American Institute of Chemical Engineers.

Sanders, R. E. (1993). *Management of Change in Chemical Plants: Learning From Case Histories*. Oxford, UK: Butterworth-Heinemann.

Savage, P. E., S. Gopalan, T. I. Mizan, C. J. Martino, and E. E. Brock (1995). "Reactions at Supercritical Conditions: Applications and Fundamentals." *AIChE Journal* 41 (July), 7, 1723-78.

Scheffler, N. E. (1996). "Inherently Safer Latex Plants." *Process Safety Progress* 15, 1 (Spring), 11–17.

Sherringron, D. C. (1991). "Polymer Supported Systems: Towards Clean Chemistry?" *Chemistry and Industry* (7 January), 15–19.

Siirola, J. J. (1995). "An Industrial Perspective on Process Synthesis." *AIChE Symposium Series* 91, 222–33.

Somerville, R. L. (1990). "Reduce Risks of Handling Liquified Toxic Gas." *Chemical Engineering Progress* (December), 64–68.

Sorensen, F., and H. J. S. Petersen (1992). "Substitution of Organic Solvents." *Staub–Reinhaltund Der Luft* 52, 113–18.

Starks, C. M. (1987). "Phase Transfer Catalysis: An Overview." *Phase Transfer Catalysis—New Chemistry, Catalysts and Applications*, September 8, 1985, American Chemical Society 190th Meeting, Chicago, IL, ed. C. M. Starks, 1–7. ACS Symposium Series No. 326, Washington, D. C.: American Chemical Society.

Starks, C. M., and C. Liotta (1978). *Phase Transfer Catalysis Principles and Techniques*. New York: Academic Press.

"Submarine!" (1992). NOVA, PBS Television Program. Boston: WGBH Educational Foundation.

Sundell, M. J., and J. H. Nasman (1993). "Anchoring Catalytic Functionality on a Polymer." *Chemtech* (December), 16–23.

Swain, A. D., and H. E. Guttmann (1983). *Handbook of Human Reliability Analysis With Emphasis on Nuclear Power Plant Applications*. NUREG/ CR-1278. Washington, DC: United States Nuclear Regulatory Commission.

Tickner, J. (1994). "The Case for Inherent Safety." *Chemistry and Industry* (3 October), 796.

Tietze, L. F. (1995). "Domino Reactions in Organic Synthesis." *Chemistry & Industry* (19 June), 453–57.

Timberlake, D. L., and R. Govind (1994). "Expert System for Solvent Substitution." *Preprints of Papers Presented at the 208th ACS National Meeting*, August 21–25, 1994, Washington, DC, 215–217. Center for Great Lakes Studies, University of Wisconsin–Milwaukee, Milwaukee, WI: Division of Environmental Chemistry, American Chemical Society.

Turney, R. D. (1990). "Designing Plants for 1990 and Beyond: Procedures for the Control of Safety, Health and Environmental Hazards in the Design of Chemical Plant." *Trans. IChemE* 68, Part B (February), 12–16.

Tyler, B. J. (1985). "Using the Mond Index to Measure Inherent Hazards." *Plant/Operations Progress* 4, 3 (July), 172–75.

Tyler, B. J., A. R. Thomas, P. Doran, and T. R. Greig (1996). "A Toxicity Hazard Index." *Chemical Health and Safety* 3 (January/February), 19–25.

Wade, D. E. (1987). "Reduction of Risks by Reduction of Toxic Material Inventory." *Proceedings of the International Symposium on Preventing Major Chemical Accidents*, February 3–5, 1987, Washington, D. C., ed. J. L. Woodward, 2.1–2.8. New York: American Institute of Chemical Engineers.

Walsh, F., and G. Mills (1993). "Electrochemical Techniques for a Cleaner Environment." *Chemistry and Industry* (2 August), 576–79.

Wells. G. L., and L. M. Rose (1986). *The Art of Chemical Process Design*, 266. Amsterdam: Elsevier.

Wilday, A. J. (1991). "The Safe Design of Chemical Plants With No Need for Pressure Relief Systems." *IChemE Symposium Series* No. 124, 243–53.

Wilkinson, M., and K. Geddes (1993). "An Award Winning Process." *Chemistry in Britain*, December, 1050–1052.

Wixom, E. D. (1995). "Building Inherent Safety into Corporate Safety, Health and Environmental Programs." *CCPS Inherently Safer Process Workshop*, May 17, 1995, Chicago, IL.

Wolfe, T. (1979). *The Right Stuff*. New York: Farrar, Straus, and Giroux.

Yoshida, T. (1987). *Safety of Reactive Chemicals*. Amsterdam: Elsevier Science Publishers, B. V.

Yoshida, T., J. Wu, F. Hosoya, H. Hatano, T. Matsuzawa, and Y. Wata (1991). "Hazard Evaluation of Dibenzoylperoxide (BPO)." *Proc. Int. Pyrotech. Semin.* 2, 17, 993–98.

9.2. Additional Reading

Ashby, J., and D. Paton (1993). "The Influence of Chemical Structure on the Extent and Sites of Carcinogenesis for 522 Rodent Carcinogens and 55 Different Human Carcinogen Exposures." *Mutation Research* 286, 3–74.

Caruana, C. M. (1996). "ChEs Seek Big Gains From Miniaturization." *Chemical Engineering Progress* 92, 4 (April), 12–19.

Crabtree, E. W., and M. M. El-Halwagi (1994). "Synthesis of Environmentally Acceptable Reactions." *Pollution Prevention Via Process and Product Modifications*, ed. M. El-Halwagi, and D. P. Petrides, 117–27. AIChE Symposium Series, 303. New York: American Institute of Chemical Engineers.

Davies, C. A., E. Freedman, D. J. Frurip, G. R. Hertel, W. H. Seaton, and D. N. Treweek (1990). *CHETAH Version 4.4: The ASTM Chemical Thermodynamic and Energy Release Evaluation Program.* 2nd Edition. Philadelphia, PA: American Society for Testing and Materials.

Drexler, K. E. (1994). "Molecular Manufacturing for the Environment." *Preprints of Papers Presented at the 208th American Chemical Society National Meeting*, August 21–25, 1994, Washington, DC, 263–265. Center for Great Lakes Studies, University of Wisconsin–Milwaukee, Milwaukee, WI: Division of Environmental Chemistry, American Chemical Society.

Dutt, S. (1996). "Safe Design of Direct Steam Injection Heaters." *5th World Congress of Chemical Engineering*, July 14–18, 1996, San Diego, CA, Paper 52c. New York: American Institute of Chemical Engineers.

Edwards, D. W., and D. Lawrence (1993). "Assessing the Inherent Safety of Chemical Process Routes: Is There a Relation between Plant Costs and Inherent Safety?" *Trans. IChemE* 71, Part B (November), 252–58.

Edwards, D. W., and D. Lawrence (1995). "Inherent Safety Assessment of Chemical Process Routes." *The 1995 IChemE Research Event, First Eur. Conf. Young Res. Chem. Eng.*, 62–64. Rugby, UK: The Institution of Chemical Engineers.

Eierman, R. G. (1995). "Improving Inherent Safety with Sealless Pumps." *Proceedings of the 29th Annual Loss Prevention Symposium*, July 31–August 2, 1995, Boston, MA, ed. E. D. Wixom and R. P. Benedetti, Paper 1e. New York: American Institute of Chemical Engineers.

Englehardt, J. D. (1993). "Pollution Prevention Technologies: A Review and Classification." *Journal of Hazardous Materials* 35, 119–50.

Englund, S. M. (1990). "The Design and Operation of Inherently Safer Chemical Plants." *American Institute of Chemical Engineers 1990 Summer National Meeting*, August 20, 1990, San Diego, CA, Session No. 43.

Englund, S. M. (1994). "Inherently Safer Plants—Practical Applications." *American Institute of Chemical Engineers 1994 Summer National Meeting*, August 14–17, 1994, Denver, CO, Paper No. 47b.

Gowland, R. T. (1995). "Applying Inherently Safer Concepts to an Acquisition Which Handles Phosgene." *Proceedings of the 29th Annual Loss Prevention Symposium*, July 31–August 2, 1995, Boston, MA, ed. E. D. Wixom and R. P. Benedetti, Paper No. 1d. New York: American Institute of Chemical Engineers.

Gygax, R. W. (1988). "Chemical Reaction Engineering for Safety." *Chemical Engineering Science* 43, 8, 1759–71.

Harris, C. (1993). "Containment Enclosures." *Prevention and Control of Accidental Releases of Hazardous Gases*, ed. V. M. Fthenakis, 404–10. New York: Van Nostrand Reinhold.

Hendershot, D. C. (1988). "Risk Reduction Alternatives for Hazardous Material Storage." *Proc. 1988 Hazardous Materials Spills Conference*, May 16–19, 1988, Chicago, Ill., 611–618. New York: American Institute of Chemical Engineers.

Hendershot, D. C. (1994). "Chemistry—The Key to Inherently Safer Manufacturing Processes." *Preprints of Papers Presented at the 208th American Chemical Society National Meeting*, August 21–25, 1994, Washington, DC, Paper No. ENVR-135. 273-5. Center for Great Lakes Studies, University of Wisconsin–Milwaukee, Milwaukee, WI: Division of Environmental Chemistry, American Chemical Society.

International Conference and Workshop on Process Safety Management and Inherently Safer Processes (1996). New York: American Institute of Chemical Engineers.

Joback, K. G. (1994). "Solvent Substitution for Pollution Prevention." *Pollution Prevention Via Process and Product Modifications*, ed. M. El-Halwagi, and D. P. Petrides, 98–103. AIChE Symposium Series 303. New York: American Institute of Chemical Engineers.

Kletz, T. A. (1983). "Inherently Safer Plant—The Concept, Its Scope and Benefits." *Loss Prevention Bulletin* 051 (June), 1–8.

Kletz, T. A. (1989). "Friendly Plants." *Chemical Engineering Progress* (July), 18–26.

Kletz, T. A. (1990). "Plants Should Be Friendly." *Safety and Loss Prevention in the Chemical and Oil Processing Industries*, October 23–27, 1989, Singapore, 423–435. IChemE Symposium Series, No 120. Rugby, Warwickshire, U. K.: The Institution of Chemical Engineers.

Kletz, T. A. (1991). "Inherently Safer Plants: An Update." *Plant/Operations Progress* 10, 2 (April), 81–84.

Kletz, T. A. (1991). "Inherently Safer Plants—Recent Progress." *IChemE Symposium Series* No. 124, 225-33.

Kletz, T. A. (1991). "Present Trends in Process Safety." *Speculations in Science and Technology, Developments in Chemical Engineering* 15, 2, 83–90.

Koch, T. A., K. R. Krause, and M. Mehdizadeh (1996). "Improved Safety Through Distributed Manufacturing of Hazardous Chemicals." *5th World Congress of Chemical Engineering*, July 14–18, 1996, San Diego, CA, Paper 52a. New York: American Institute of Chemical Engineers.

Lawrence, D., and D. W. Edwards (1994). "Inherent Safety Assessment of Chemical Process Routes by Expert Judgement." *The 1994 IChemE Research Event*, 886–88.

Lawrence, D., D. W. Edwards, and A. G. Rushton (1993). "Quantifying Inherent Safety." *Proc. Inst. Mech. Eng.* 4, 1–8.

Leveson, N. G. (1995). *Safeware: System Safety and Computers*. Reading, MA: Addison-Wesley Publishing Company.

Lubineau, A. (1996). "Making a Splash in Synthesis: Water As a Solvent." *Chemistry and Industry* (19 February), 123–26.

Lutz, W. K. (1996). "Advancing Inherent Safety into Methodology." *5th World Congress of Chemical Engineering*, July 14–18, 1996, San Diego, CA, Paper 50a. New York: American Institute of Chemical Engineers.

Mansfield, D. P., and K. Cassidy (1994). "Inherently Safer Approaches to Plant Design." *IChemE Symposium Series* 134, 285–99.

McCarthy, A. J., J. M. Ditz, and P. M. Geren (1996). "Inherently Safer Design Review of a New Alkylation Process." *5th World Congress of Chemical Engineering*, July 14–18, 1996, San Diego, CA, Paper 52b. New York: American Institute of Chemical Engineers.

Melhem, G. A. (1993). "Hazard Reduction Benefits From Reduced Storage Temperature of Pressurized Liquids." *Prevention and Control of Accidental Releases of Hazardous Gases*, ed. V. M. Fthenakis, 411–37. New York: Van Nostrand Reinhold.

Mittelman, A., and D. Lin (1995). "Inherently Safer Chemistry." *Chemistry & Industry* (4 September), 694–96.

Orrell, W., and J. Cryan (1987). "Getting Rid of the Hazard." *The Chemical Engineer* (August), 14–15.

Penteado, F. D., and A. R. Ciric (1996). "An MINLP Approach for Safe Process Plant Layout." *Ind. Eng. Chem. Res.* 35, 1354–61.

Rushton, A. G., D. W. Edwards, and D. Lawrence (1994). "Inherent Safety and Computer Aided Design." *Trans. IChemE* 72, Part B (May), 83–87.

Singh, J. (1993). "Assessing Semi-Batch Reaction Hazards." *The Chemical Engineer*, 537 (February 5), 21, 23–25.

Snyder, P. G. (1996). "Inherently Safe(r) Plant Design." *1996 Process Plant Safety Symposium*, Volume 1, April 1–2, 1996, Houston, TX, ed. H. Cullingford, 203–215. Houston, TX: South Texas Section of the American Institute of Chemical Engineers.

Speechly, D., R. E. Thornton, and W. A. Woods (1979). "Principles of Total Containment System Design." *North Western Branch Papers No. 2, Institution of Chemical Engineers*, 7.1-7.21.

Steensma, M., and K. R. Westererp (1988). "Thermally Safe Operation of a Cooled Semibatch Reactor. Slow Liquid–Liquid Reactions." *Chemical Engineering Science* 43, 8, 2125–32.

Steensma, M., and K. R. Westererp (1990). "Thermally Safe Operation of a Semibatch Reactor for Liquid–Liquid Reactions. Slow Reactions." *Ind. Eng. Chem. Res.* 29, 1259–70.

Velo, E., C. M. Bosch, and F. Recasens (1996). "Thermal Safety of Batch Reactors and Storage Tanks. Development and Validation of Runaway Boundaries." *Ind. Eng. Chem. Res.* 35, 1288-99.

Verwijs, J. W., H. van den Berg, and K. R. Westerterp (1996). "Startup Strategy Design and Safeguarding of Industrial Adiabatic Tubular Reactor Systems." *AIChE Journal* 42, 2 (February), 503–15.

Weirick, M. L., S. M. Farquhar, and B. P. Chismar (1994). "Spill Containment and Destruction of a Reactive, Volatile Chemical." *Process Safety Progress* 13, 2 (April), 69–71.

Whiting, M. J. L. (1992). "The Benefits of Process Intensification for Caro's Acid Production." *Trans. IChemE* 70 (March), 195–96.

Windhorst, J. C. A. (1995). "Application of Inherently Safe Design Concepts, Fitness for Use and Risk Driven Design Process Safety Standards to an LPG Project." *Loss Prevention and Safety Promotion in the Process Industries*, ed. J. J. Mewis, H. J. Pasman, and E. E. De Rademacker, 543–54. Amsterdam: Elsevier Science B. V.

Index